REMEMBERING ROSTOCK 1972-1990

The Proceedings of the
Ahrenshoop Symposium
of the Brown-Rostock Exchange
March 1989

edited by
Hans-Joachim Bernhard and Duncan Smith

UNIVERSITY
PRESS OF
AMERICA

Lanham • New York • London

University Press of America®, Inc.
4720 Boston Way
Lanham, Maryland 20706

3 Henrietta Street
London WC2E 8LU England

Printed in the United States of America
British Cataloging in Publication Information Available

Library of Congress Cataloging-in-Publication Data

Ahrenshoop Symposium of the Brown-Rostock Exchange (1989)
Remembering Rostock, 1972-1990 : the proceedings of the Ahrenshoop Symposium of the Brown-Rostock Exchange, March 1989 / edited by Hans-Joachim Bernhard and Duncan Smith.
p. cm.
1. United States—Relations—Germany (East)—Congresses.
2. Germany (East)—Relations—United States—Congresses.
3. Exchanges, Literary and scientific—United States—Congresses. 4. Exchanges, Literary and scientific—Germany (East)—Congresses.
5. Brown University—Congresses.
6. Universität Rostock—Congresses.
I. Bernhard, Hans Joachim, 1929- .
II. Smith, Duncan. III. Title.
E183.8.G35A36 1989
303.48' 2730431—dc20 91-14795 CIP

ISBN 0-8191-8370-9 (hardcover, alk. paper)

The paper used in this publication meets the minimum requirements of American National Standard for Information Sciences—Permanence of Paper for Printed Library Materials, ANSI Z39.48–1984.

Rostock 1989

This volume is dedicated to

Dr. Jerome Ottmar

with appreciation for his generous support

TABLE OF CONTENTS

Introduction

Today we find ourselves in the midst of a process of profound changes. Having been moved only two hundred years ago by the great aims of the French Revolution and the hope for "liberté, égalité, fraternité" - humanity, at the end of this millenium, is facing the question whether our history is coming to an end. This question became part of our daily agenda because of the threat of nuclear self-annihilation and the self-made looming ecological catastrophe that we have so far not seriously attempted to allay. No less a danger to humanity is the rapid decay of moral values and the impending triumph of consumerism and egoism over socially responsible behaviour and human solidarity.

Scholars and scientists have to assume a special kind of responsiblity in such a situation. It is a special kind of responsibility for all of those committed to the advancement of the humanities and sciences to speak out against ignorance and against the dissemination of hate images and against ecologically irresponsible behaviour and against the waste of resources. Academics have a duty to uphold the principle of objectivity in the examination of ideas, to dispute without rancour and to advance the cause of the socially deprived in society.

The authors of this preface were guided by these principles in their efforts to bring together academic colleagues from Brown University and from the University of Rostock. These efforts began after their first meeting in 1972 and despite many difficulties came to fruition in 1979 with the signing of an agreement. On the tenth anniversary of the agreement, the occasion was marked by a symposium attended by colleagues and students of both universities and held in Ahrenshoop, March 16-18, 1989. The title of the symposium was "The Benefits of Dialogue and Academic Cooperation -Ten Year Exchange Program Brown University - Wilhelm Pieck University." As the title indicates, this conference was not an academic exchange in specific fields but focused upon the directions of former and future cooperation. This volume summarizes some of the contributions to this symposium.

Joint endeavors in the humanities are surveyed in this article by the Germanists Duncan Smith and Hans-Joachim Bernhard. They are indebted to the concept of *Gedächtnisarbeit* (the task of rememberance), a term coined by the German humanist Johann Gottfried Herder. By the task of rememberance we mean an examination of the past through literature, learning its lessons and thereby forearming literature against the menaces of our own times.

The contribution by Karl Heinz Jügelt, head librarian of the University of Rostock Library, is a convincing example of the importance of the exchange of information. The contributions made by J. Kestin and Jürgen Millat discuss in specific terms the progress made by research groups in both countries in the field of chemistry and also the benefits of cooperation for the development of scientific research in general. It becomes apparent that this scientific cooperation encompasses more than just the groups of researchers at each university. The mutual benefit of growing relations between the people of two countries ensured by the agreement played an important role during the symposium. For this reason the volume includes contributions by younger academics as well. Bernd-Peter Smollich (Medical School), Volker Ahnfeld (English) and Joseph Polman (Education) talk about what the opportunity to study in the other country meant to them not only academically but personally.

Thus, this volume - which is only a selection from the many contributions to the symposium - illustrates clearly the benefits obtained by the agreement between the universities for the development of academic research, for the personal development of those involved, especially for the younger researchers and the students. Without exaggeration, one can say that the agreement has been a constituent for the growth of trust between our peoples, a step toward mutual understanding and the growth of a peaceful and just world.

At this point, we want to thank all the people who helped to realize and to promote this project. Our special thanks are directed to Dr. Jerome Ottmar Senior who supported our cooperation steadfastly from the very beginning. Since the unification of the two Germanies in 1990, this volume has become a documentation of an historic program which in the most difficult times promoted peaceful dialogue and then, as now, the arduous task of cross-cultural cooperation.

Hans-Joachim Bernhard Duncan Smith

Rostock, January 1990

Professor Hans-Joachim Bernhard and Professor Duncan Smith at the September 1979 Contract Signing

Howard Swearer and Wolfgang Brauer in Rostock, 1981

Rektor Brauer and Dr. Jerome Ottmar in Rostock, 1982

The "Task of Rememberance" in Recent German Literature of German Speaking Countries

Hans-Joachim Bernhard

An examination of recent literature reveals that the questions and challenges at the end of the century take on a more comprehensive character and must necessarily do so. Any analysis of literature must take this into account to an increasing degree but without ignoring specific national issues and characteristics.

When planning the contribution to the colloquium in Ahrenshoop I started from the assumption that the complex motif of Remembering in post-War German literature - which turns out to be part of the critical examination of the Fascist period - had not, unfortunately, lost its importance. Its continuing relevance is also due to the new approaches it has received at the hands of GDR authors such as Christoph Hein and Alfred Wellm - to be discussed in greater detail later - and can clearly be seen in their work. West European literature also exemplifies this motif of remembering through an examination of the most recent past with respect, for example, to the expectations, illusions and outcomes of the students' movement of '68.

It was Siegfried Lenz who, speaking on the occasion of the award of the Galinski Foundation Prize, referred to Herder's coinage "the task of rememberance." The task of rememberance is understood as the process of searching, questioning, and appropriating of "whatever the testing agenda confronts us with." Lenz went on to say, "For me, there is no doubt that once we disown memory, once we abandon all the foregone conclusions to which memory obliges us, we are on the path of political and moral betrayal which is also a betrayal of our humanity." One can go further, memory can be a defining characteristic of what it means to be truly human.

To be human entails a willingness to accept the burden of rememberance...The task of rememberance is to search, to question, to appropriate, to take the past on board no matter what the cost. We have to be willing to do that, for "only he who acts may hope." Those words were uttered not only in protest against the process of a selective rememberance in which the ghosts of the demonic Nazi past were tucked away but also to a process applying to the more recent past, in the GDR as elsewhere.

Situated as we are today in the midst of huge social upheavals, this theme of rememberance becomes endowed with an unforseen topicality and relevance. I am concerned with the dialectics of the appearance, process and development which is the purpose of the task of rememberance and to make a plea for the informed viewpoint which is the arch-enemy of selective rememberance. A memory which chants only a litany of successes and self-congratulations as has been the case in the GDR for decades will end finally in a rupture, will end, as Lenz put it, in the abandonment of political morality, the betrayal of humanity. Literature, at its best, has always borne witness to the task of rememberance and contributed thereto.

In Heinrich Böll's novel *Haus Ohne Hüter*, the protagonist Nella Bach meets Lieutenant Gäseler, the former immediate supervisor of her husband, the poet Rai, and responsible for his death. The utter small-mindedness and superficiality of Gäseler's character convinces Nella to abandon her scheme of vengeance. However, finally goaded by Gäseler's endless boasting of his exploits and drivel about Erwin - he is referring to Rommel - she objects, "Oh yes, you forgot the war, but remember the general's Christian name." This is a key phrase in the novel and is important not only for the interpretation of this work but far beyond it. This is a very early reference - *Haus Ohne Hüter* was published in 1954 - to the disjuncture between what happened, what was experienced, and subsequent societal developments. This question of remembering and forgetting has been taken up as a running theme in literature and not only by Böll. It is a sifting of history together with the moral categories involved in rememberance and an exposure of a convenient one-eyed view of history.

Recent works by Austrian writers provide exemplary illustration of the rigorous "sifting" and "tunnel-vision" involved in the process of remembering. I refer to the novel *Die Kleine Figur meines Vaters* (My Pathetic Father) by Peter Henisch. First published in 1975 and re-issued in a revised version in 1987, the novel takes place among the series of so-called father books of the seventies such as those by Christoph Meckel and Peter Härtling and confronts most questions about our relationship to reality.

I intend to highlight such questions in relation to my topic in what follows and ask for your forbearance that I do not deal with the highly interesting structural features of the novel such as the interplay of fiction and documentary material, the role of the narrator and the placement of individual episodes and so on.

For Henisch, the past to be examined is the period of fascist rule. This is not of course, to clarify our knowledge concerning "facts": the sequence of events, conflicts and processes etc., but in order to make an existential enquiry. Narrator and author are established as the same person in need of gaining a clearer understanding of his relationship to his father, an impassioned photographer who was acknowledged to be "the best war-correspondent of the entire German *Wehrmacht*." In this way he is able to take stock, not of his own thoughts and deeds to date, but of his father's life. He needs to know who his father is in order to answer, "Who am I?" Walter Henisch, the father, records on tape the story of his life to which the author adds tales told him by his father, his own memories and letters to his father.

In one of these letters he writes, "Behind another, one discovers oneself. Perhaps true, but not, I hope, in the sense of an unchangeable identity. I don't want to follow your footsteps. Do you understand? I don't want to relive your life." For a long time, painfully long, the reader is regaled by the father's carefree, lively stories of life in the Third Reich free from authorial commentary; stories bereft of any historical and moral dimension until finally the son bursts out, "Why did you go along with that?" At some point in one's life there has to be a break after which a person has to have some responsibility for himself. The question addressed here is not simply one of identity with another in a historical or a bodily sense, nor one of generational succession but of the establishment of true individuality, the hard-won identity of a distinctive character.

Taking stock, personal identity, the breaking point where responsibility begins, these are the building blocks of my intentionally discursive theme.

With the individual opus as a starting point these concepts allow for a somewhat wider framework of discussion in which later I would like to include a consideration of Michael Scharang's *Harry* and Alfred Wellm's *Morisco*, all the while continuing to limit my focus to the special situation of the artist in society. We must forego any consideration of the psycho-sociological debate surrounding the complex remembering/repressing process connected mainly with the name of Mitscherlich, an impossible inclusion in this essay.

Peter Henisch is aware that he will never discover his true identity as a prerequisite for his future work unless he is freed from all traces of his father's attitude. The father, on the other hand, is silent about all the historical and moral implications of his experience. When asked about what he knew about crimes committed against the Jews, officially or unofficially, his father resorts to the fable of the three wise monkeys: "See no evil, hear no evil, speak no evil, my father boasted."

Equally, Peter Henisch is aware that to cut off ties with his father with harsh moralizing judgements is also to block his own path to discovering his true identity. He must continue to find a way to understand his father lest categorically rejecting him he leaves undiscovered that within himself he had supposed he had rejected. Thus, the attempt to reach his father is an expression of human self-respect. He tries to understand and is sorry for him. As a starting point he is helped by the self-knowledge that he has inherited some of his father's attitudes toward the world.

"As a young press photographer I was less interested in who against whom and the whys and the wherefors. What captured my imagination and very often even fascinated me were the EVENTS IN AND FOR THEMSELVES. The events happened before my eyes, my camera, they were nothing but motifs for me. Even when I was in the midst of things, I always felt a certain detachment, always felt, in a sense, outside of what was happening around me."

This attitude is not a very novel one at all. It is an everyday manifestation of a basic philosophical aestheticism: "Life's a rich paegent, etc." It is found unadorned in the literature of the First World War, for example, in Ernst Jünger's diary *In Stahlgewittern* where he experiences the massacre of prisoners only on an aesthetic level or, as he puts it, witnessed "from a box seat in the theater."

Thus a sphere of activity is created which permits no room for moral judgement. Similarly, historical judgement is suspended by reference to Fate which is simply what it is and one is at its mercy. The flight into pure Art without responsibility must be contrasted with the basic philosophical aestheticism of the photographer Henisch for whom it is a mere emotional survival kit. Still, he would much rather like to see himself as a Schwejk.

It is to the credit of Henisch's novel that its investigation illuminates the causes that lead to the attitude of the father. There is first of all the effective call to duty and obligation. Secondly, there is the important experience during the war in France of being regarded as "so much bigger and stronger" for a physically small man who was constantly teased about his size as a child. Thirdly, there is the addiction to the "forward euphoria" as well as the temptations of a successful career and flattery from the Nazis who had a "nose for talent."

The father's account gives the son a chance to "distance himself from undoubtedly important character traits." He is aware of the danger inherent in the fact that everything that surrounds him, that exists and that happens assumes for him not motifs but by virtue of his work the shape of "raw material", that is, separated from a judgeable context. The sensitivity of the son who puts a concerned approach in the place of cold rejection finds reward, not only of greater certainty about himself, but also the benefit of pushing ahead with an account of a life approaching its end. The intention of the father in the little bit of time that he has left to stop"this mendacious snap-shooting" that was asked of him his whole life and "at least now to start being truthful and honest in his work" does not turn into a cheap happy end. On the contrary, it leads to knowledge, knowledge gained from breaking up the crusts of life. The novel illustrates this process with an artistically convincing slowness, uncovering the remnants of moral substance of an existence deformed by the sociopolitical conditions.

The aim of our work here is to make visible common ground as well as differences in order to learn from each other. It I take the liberty of drawing your attention to the context at this point, I do so in order to show how the literature of German speaking countries as well as all other literatures see themselves faced with the task of trying to give their best in the attempt to find answers to the great challenges of our time. This context includes the remembering of past experiences as well as the emphasizing of moral and historical elements of these experiences within the literary process and its new impulses. I think that the presence or absence of moral and historical elements within the process of remembering our near and nearest history will play a big role in this attempt. The challenge at the turn of the century will turn out to be exactly the challenge of the moral and historical dimension in our thoughts and actions. The attempts to defame reason and moral judgement have for a long time not just been isolated attempts and the "anything-goes" euphoria surrounding postmodernism will turn out to be a touchstone of historical and political identity, especially a touchstone for the identity of the writer.

We are facing a historically and politically new situation. The West German post war generation of authors as represented as e.g. by Böll, kept awake the moral essence of their war memories. Denying historical implications, this generation nevertheless did not avoid them when structures of restoration and recurrence became apparent.

Any discussion about individual experiences of the years of Nazi rule and the war that could possibly cause identification, was being repressed in the GDR. These experiences as well as moral judgement seemed to be resolved in the sense of the Hegelian concept *Aufhebung* within the newly created social structures which do not allow for a return of the fascist disaster. Nevertheless a gap became noticeable. The individual reflection on the past grounded in personal experience should not just be replaced by logical methods of thinking that impart historical experience convincingly but abstractly. Fühman's own development from the poem about Stalingrad to the essay on Trakl is proof enough of that. The questions that Christa Wolf touched on in *Kindheitsmuster* (Patterns of a Childhood) required an answer: which remnants of thinking and behavior patterns continue to exist in a hidden form? The literature of German speaking countries is evidence that the subject of the fascist past cannot be put aside, especially not in a place where it is more and more common openly to justify the atrocities of fascism. The literature of the eighties offers quite a few examples of individual assessments that take as their subject the post-war development.

One example of great importance for this development is in my opinion Christoph Hein's *Die Ritter der Tafelrunde* (The Knights of the Round Table); an attempt to make an assessment by taking the Arthurian legend as its initial model. This assessment concerns the way of life under newly developed social structures, i.e., socialism, but that does not mean that there are not comparable attempts to be found in Austrian or West German literature.

The sociopolitical movements of the sixties and early seventies found their hopes reflected in the literature of Austria and West Germany. I consider Michael Schareng's text, *Harry, eine Abrechnung,* a document of high artistic value within the literary examination of unfulfilled expectations and of consequences for a way of writing that is committed to the idea that reason can influence society. One fundamental and enticing idea determining the structure of the relatively short text offers the reader material of considerable philosophical depth and much potential relevance to the present historical situation. A young man of 25 years is asked by a writer to come to the local inn for purposes of revenge. Their relationship is more or less coincidental and is grounded mostly in common circumstances. The young man was a friend of the writer's son, Andreas. When Andreas, who is now a paraplegic, had a car accident, the young man was also in the car. On the other hand, they also have a kind of spiritual relationship. The young man is the son of Charly Traktor. The writer never gets a chance to speak, since the text is a monologue addressing him. The author's questions can only be deduced from the utterances of the speaker. Thus, it could be said that the text is multi-layered, which allows for a plenitude of associations.

Within the framework of this essay, we can of course, only discuss one facet that is important to our analysis of this fully developed aesthetic design. What is of interest to us is the mediated way in which the author comes to terms with his own life. In presenting the assessment of a young life, the narrator who is the young man, is also included in the act of reckoning. He assessed his life in order to demonstrate how useless it would be to attempt to write a book about the narrator: "I'll give you one piece of advice: don't write this book." This is the first sentence of the text. The narrator does not want his image to be idealized like the one of the worker Charly because he does not want important things to be suppressed. Charly, the father, would stuff every single page into the mouth of the writer, his Creator, if he could only get hold of him. The father may have learned "how one can change things but he relies on methods that have been buried long ago." The father is still "a courageous labor unionist" whom nobody can buy, but he no longer has any bite. It sounds incredibly presumptuous when the son declares that he already understands more than his father will ever understand. This point marks only a radical renunciation of illusions and the insight into the mechanisms of the world he has to cope with. What is being negotiated here is not simply the rejection of fathers determined by the generation gap but also the recapitulation of an evaluative remembering that is also a reckoning revealing its historical and moral substance. The narrator declares his support for his father when he says, "I do not distance myself from my father and certainly not from his cause." His statement revolves around the continuity of values that will be damaged if pursued by means that have been eliminated by progress long ago.

Scharang found a powerful metaphor for the attitude that is being solicited. the narrator grew up bilingual. There is the language he uses to talk with his mother and the one he uses to talk to strangers. He talks extensively about his experiences with strangers. He tells everything: from the times he and the other children played each other off in kindergarten to the time when he murdered his girlfriend Evi who knew too much about a high-ranking person. When his mother, concerned about his advancement in society, attempts to force him into the service of an influential landowner, he can only find one word for it, "betrayal". He loses his bilingualism. "The language I could trust" did not exist any longer. He discovers his identity within his own language. He starts to write. In contrast to his father he is able to express himself. We encounter him having no illusions and as knowing his escape will not lead to paradise but not fully certain whether he will not "betray" his own identity, "his life" one day.

The subtitle, "Resumé of a life," would also be appropriate for Alfred Wellm's novel *Morisco* It is even more complicated to discuss our problematic since we are dealing here with a highly differentiated text in terms of narrative structure. It will not be possible within the framework of this essay to discuss the author in detail, nor to offer in-depth evidence that we are discussing here one of the most important works of narrative prose in GDR literature of the 1980's. The novel can be placed within the tradition of the *Weltanshaungsroman* and indicates the comprehensive character of the problematic discussed in this essay. The motif of the task of rememberance is taken up in different textual forms, whether it be a text mainly shaped by documentary material, a narrative or, as in this case, an expansive novel.

We started our investigation with the task of taking stock, the attempt to preserve one's personal identity, the break-point in one's life that provokes responsibility. How is all of this represented in *Morisco*? Andreas lives in a room in a tower of a castle in the North of the GDR. He can look back on a fast career, on fifteen successful years as a construction supervisor and on a position as chief architect of a major city. Now he is supervising the restoration of the castle he lives in. He allows one of his colleagues, Matthias, to have a "look into his life" Lenk tells him what happened to him, how his 18-year-long marriage with his wife Anna broke apart and why he decided to make a new start in his professional and social life. This is the narrative situation. Nevertheless, Matthias does not play the part of the father-confessor. What is at stake here is not the granting of the absolution but the sharing of thoughts and emotions. What is unfolding in front of us is the weighing of decisions, things taken for granted as well as objections. *Morisco* also appears to me to be a good example of how the relationship of the text to the reader as it is specific to GDR literature manifests itself in the narrative form. The résumé of a life asks to be put into question, indeed, it demands it. It makes it easier to test it since it does not attempt to cover up anything.

The four parts of the novel parallel the main stages of a development. The first part leads us back to the student days and illustrates goals in life. The second part highlights the contradictions caused by his start as a construction supervisor and his fast but not forced career. The third part centers around the crisis and its solution based on Lenk's own initiative. The fourth part unfolds the possibilities and difficulties of a new start. When Lenk and his girlfriend and future wife, Anna, find each other, they also find themselves in accordance with a society that is led by the insights of the Communist Manifesto. They discuss its statements and Anna, concerned, inquires about the possible constraints and obligations:

"And there is surely no obligation?" "Not in the sense that you have to act against your own insights, against your own will." - "But in that case you have to know about everything." - "Yes, about everything." - "Also about everything happening within a country?" - "Of course. Since it is the highest form of democracy, nothing will be withheld from the people." - "And that will happen?" - "It certainly will happen," I said."

Lenk does not want the "peace of being content." That attitude, of course, forces him into the conflict between being content with the power relations, with the shallow life as a consequence of insidious corruption and being content out of necessity. Of course, the possibilities for building cities like Heliantha 1 and 2 as they take shape on Lenk's drawing-board at night do not yet exist.

It can be neither discussed nor proven within the framework of this essay how convincingly Wellm describes the mechanisms which determine Lenk's and Anna's future actions. Comrade Sarazin, the main person responsible for the construction of the city, shows a certain liking for the energetic construction supervisor. But what begins with a harmless invitation to a party at the lake and the insistence that he take a bigger apartment ends in "the pact with the devil" for the sake of the greater cause. The supply of arguments for self-justification is taken from an old but well-known arsenal: If I don't do it, someone else will do it - are Lenk's thoughts when the mayor asks him to design a house for a large amount of money. A life of "double-existence" begins, a "second honesty" is spreading within him. He forfeits his solitude. Solitude stands for identity in Wellm's work. While having a conversation about literature with the wife of a mayor who feels "left alone" by the author of the book, Lenk already remarks, "The greater the loneliness (*Einsamkeit* - Onesomeness) within a person, the stronger the power in him for the common good."

This becomes even more obvious when Lenk justifies to himself why he and Anna seek the company of Josef, the sculptor and graphic artist. What impressed them was, "his loneliness. He was himself...Said what he thought."

An event concerning Josef finally causes the breaking point in life where responsibility begins. Anna and Lenk manage to convince the otherwise secluded sculptor to participate in a competition for a monument in the central square. Not quite without help from his friends, his design is chosen. Lenk finds the finished work captivating. A protest disguised as coming from the workers against the nudity of the figures and the missing optimism of the monument causes its rejection, desired by the higher authorities. When Lenk is asked to make the final decision about the monument's fate, he changes sides and adopts the opponents' view. As in Scharang's text, the word that comes to mind at this moment is betrayal. Betrayal because it designates the surrender of one's self for Wellm. "...that we betrayed everything, yes, everything and on top of this our irrationality, our truths and convictions."

The "we" refers to Sazarin, the father of Lenk's career, who, as Lenk believes, loves him like a son. Here again we find a similarity to Scharang's text since Lenk will never be able "to despise Sazarin wholeheartedly." He knows all too well how torn up the older one is inside, and how unselfish his efforts for a better cause have been. Nevertheless, he distances himself from him.

The novella that is told in the fourth part is the story of a young but unfulfilled love. The passion that upholds this love evokes the question in the narrator: "But does passion - however it may appear - always lead to a destructive end? No, it is always a new beginning."

Finally, let me give you three instances that exemplify Wellm's fine performance as a narrator. On the one hand there is the inner necessity that finds its multi-faceted expression in the representation of the correlation between Lenk's relationship to his wife Anna and his actions in public. On the other hand it is impressive how the narrator does justice to his characters. The merciless self-questioning of the narrator does not allow for any satirical punishment, any conviction of others that would let him evade responsibility.

A last point to be discussed is the epic frame. At night in his tower Lenk imagines the castle crowded with figures from the Renaissance, imagines their thoughts and actions and imagines himself to be one of them. These references, though sometimes overemphasized, direct the thoughts of the reader toward the substantial, the paradigmatic in the course of human life as presented. In this context it is justified to reflect on whether and to what degree humans have changed since those days when the castle was built. Criticism granted Wellm's "moral rigor" concerning "lived history and consciously experienced life." I think we can also find this kind of rigor in Henisch and Scharang. Art needs this rigor if it wants to meet the challenges of our days. Art does not need it as a prescription or as a legal code, but rather in the way the narrator comments on the work of the sculptor Josef: "hence, not a result but a process....a thought still in the process of creation. Art could not do more." I do think, however, that is more than enough.

Translated by Katrin Schröter

Reading lists, literary canons and cultural literacy

William Crossgrove

When thinking about what to say at a colloqium to commemmorate the Brown-Rostock exchange, I was reminded of an informal talk given in the fall of 1988 at the German House on the Brown campus by Georg Moll of the History Section at the Wilhelm Pieck University, who was then visiting Brown as part of our exchange. He spoke on *Tradition und Erbe in der Gewissenschaft der DDR* and, simplifying for the benefit of American students, clearly laid out the distinction between *Erbe,* the totality of cultural transmission, and *Tradition*, those parts of the *Erbe* that are selected for emphasis and for emulation by subsequent generations. The *Erbe-Tradition* discussion has of course been a central theme of Marxist thinking in this century, but a comparable discussion in the laissez-faire cultural world of the United States has usually occurred under the guise of other issues. Without much warning, such a discussion has now broken out in full force, through it still occurs with a somewhat different terminology.

Indeed, the terms *Erbe* and *Tradition* as used in Marxist discussions are not easily translatable into English. The English term "tradition" often comes closer to the meaning of *Erbe* than to *Tradition*, that is, it often refers to the totality of a cultural heritage. The selection of literary texts to be emphasized and used in training new generations of students is usually justified with reference to aesthetic criteria. The textx to be included in this literary canon are those that are "best" by aesthetic norms that are popularly thought to transcend historical change, to reflect lasting values. Most scholars do not defend such a simplistic view of ahistorical categories, yet the practical decisions are often made as if transcendent aesthetic criteria still existed.

Among the most discussed books of 1987 were two that are unlike most bestsellers both in terms of their subject matter and their authors. The first of these was The Closing of the American Mind by Allan Bloom, a professor of philosophy at the University of Chicago (Bloom 1987). It is only a slight exaggeration to say that his book describes a continuous decline in the level of philosophical discourse from Socrates to the present, but it is no exaggeration to observe that the author thinks that the pace of this decline accelerated sharply in America in the nineteen-sixties with the rise of the "New Left." He also readily identifies the source of the precipitous decline. It is allegedly a legacy of moral relativism that has its modern origins in the philopsophy of Nietzsche and the social theories of Max Weber. Among the chapters headings are "The German Connection," "The Nietzscheanization of the Left or Vice Versa," and "From Socarates' Aplology to Heidegger's Rektoratrede." Bloom thinks that the proper road to overcoming this moral relativism is to return to the Great Books of Western Civilization where we can find eternal verities that are not subject to the vicissitudes of the latest fashion to sweep our culture.

Bloom sowed his seed on fertile ground. Secretary of Education Wiliam Bennet had been using a similar theme to make headlines even before he became Secretary of Education. The great hope for a renewal of American civilization was to be sought by turning the clock back to a time when women's literature, literature of the Third World, and literature of American minorities did not exist, at least in the mind of white male intellectuals. Bloom's book was widely hailed by thoses who took comofrt in his diagnosis of the failings of modern society, and sales were astonishingly good for a book by a philosophy professor. Ironically, one has to wonder how many of the supporters of a return to the Great Books actually read Bloom's book. It is not easy bedtime reading. I do not agree with his premises or his conclusions, but his arguments are serious and well presented.

In the same year, E.D. Hirsch, Jr., a professor of English at the University of Virginia published Cultural Literacy, an argument in favor of the identification and propagation of a body of basic factual information that is shared by most educated participants in American national culture (Hirsch 1987). His starting point was the claim that there is a body of information in any given culture that must be understood by all readers who want to participate fully in that culture. He reviews the literature on the pedagogy of reading and argues that children who do not acquire basic cultural information will remain at a disadvantage by comparison with those who are full participants in national culture. Hirsch advocates a program of targeted teaching of basic information in the interests of democratic enlightenment. With the help of several colleagues, he put together a list of about 5000 basic concepts that American high school graduates ought to recognize at least vaguely if they are to be able to read newspapers or watch television with some degree of understanding.

Hirsch acknowledges that his list overwhelmingly reflects the dominance of white male interests, but he claims that this is an empirical finding about what Americans need to know in order to be culturally literate. Changes will occur only after popular culture has been sufficiently permeated by new values. Despite Hirsch's attempts to explain this point and to argue that superficial factual knowledge of cultural literacy needs to be supplemented with in-depth study in most subjects, he is often linked with Bloom (see, e.g., Swaffer 1989, p.57). In some circles, Bennet, Bloom and Hirsch have become a kind of troika, either of inspirational leaders or of voices of reaction, depending on how people view what they advocate, or are thought to advocate.

Before discussing the broader issues of canonical texts and cultural literacy, it is useful to examine some more specific implications of Hirsch's findings for those who teach German in the United States. He assumes that each national culture would have its own list, since the national culture is the contxt in which media operate and general discussion of cultural matters occurs. His list provides us with food for thought about German-American cultural relations. I tried to ascertain how many of the 5000 items pertain more or less directly to German speaking countries. By my count, there are 116: less than one in 40, slightly over 2.3% of the total. Forty-one of these are names, inculding nine from music (Bach, Beethoven, Haydn, Mendelssohn, Mozart, Schubert, J. Strauss, R. Strauss, and Wagner) and six fom the natural sciences (Einstein, Kepler, and Leibnitz and three who are included only because something is named after them: the Heidelberg effect, Hertz and Planck's constant). There are five philosophers (Hegel, Kant, Marx,

Nietzsche and Wittgenstein), five Nazi leaders (Eichmann, Goebbels, Göring, Himmler and Hitler), and five names from German literature (Goethe, Kafka, Thomas Mann, the Grimm Brothers and Schiller). Two of the eighteen geographical names are concentration camp sites, namey Dachau and Auschwitz. Among the eleven named texts, five are fairy tales or legends: Hansel and Gretal, Little Red Riding Hood, The Pied Piper of Hamelin, Rumpelstiltskin, and Snow White. *Faust* and Kafka's *Metamorphosis* represent German literature, while the *Communist Manifesto* and *Das Kapital* are included along with Silent Night and *Mein Kampf*.

This is not exactly a cause for celebration for Germanists, but before we conclude that Hirsch knows too little about German culture, it is worthwhile to examine the first major response from critics of his conservative bias. A collection of essays by leading representatives of groups that have commonly been excluded from the mainstram of American culture concludes with a list of 600 entries proposed as additions to the list created by Hirsch and his collaborators (Simonson and Walker 1988, pp. 191-200). The purpose of the new list is to promote mult-cultural literacy, awareness of traditions other than the old mainstream one. It contains something of the prescriptive component ostensibly lacking in the Hirsch list. The number of entries pertaining to German speaking cultures is 14, the same 2.3% as we found with the original list. The eight names among these fourteen items refer, along with Dietrich Bonhoeffer, Bertolt Brecht, Günter Grass and Rainer Maria Rilke, to four people who became better known in the Anglo-Saxon world than in their countries of origin (Alfred Adler, Anna Freud, Eric Fromm and Paul Tillich). Among the geographic names are two included because they are associated with World War II atrocities, namely Dresden and Buchenwald.

In other words, the alternative list does not differ markedly from the original list in terms of references to German speaking cultures.

The context in which we pursue the study of German in the United States thus is one in which our students do not even begin to approach literacy in any German national culture. The situation described above is not unique for students of German studies. Colleagues in Slavic languages face even more difficult problems with American students, those in Romance languages appear to be only marginally better off, and colleagues teaching non-European languages can presuppose almost nothing. It is also not self-evident that the proportion of information about European matters is inappropriately small for American students. But seeing it reflected in data about what American students might be expected to know is sobering.

There are many questions about the value of the kind of simple factual knowledge advocated by Hirsch as a basis for cultural understanding (see Colomb 1989a, Colomb 1989b and Hirsch 1989 for a recent debate on some of the basic controversies), yet the resonance that his work has found suggests that he has raised an important issue. He has opened a discussion about our *Erbe*, about the transmission of basic knowledge necessary to understand the culture of the United States. I think that there is a lot more to be said in this discussion, and it ofers us a unique opportunity for cooperation in this sense: Germanists working in the United States might want to think seriously about what American students need to know about German culture, in order to have German studies be comprehensible to them, and Americanists working in German speaking countries might want to consider the same points about the cultural literacy of their students. It suggests a possible area of cooperation between our two institutions.

Hirsch now heads an organization called the Cultural Literacy Foundation that is working to refine lists and to encourage schools to adopt his ideas. The list now forms the basis of a more ambitious Dictionary of Cultural Literacy that organizes over 7000 items into 23 categories ranging from "The Bible" to "Technology," with entries alphabetically arranged in each category and a general alphabetical index at the end (Hirsch, Kett and Trefil 1988). For each entry there is a brief explanation, presumably the essential knowledge that would serve either for cocktail party chit-chat or for understanding references in newspapers, magazines and public debates. Hirsch's effort has also received anothr kind of acknowledgement that it resonates in the American public. A book has appeared that allows people to test their cultural literacy, presumably giving them the opportunity to feel good if they can pass the tests and inadequate if they fail.

It should be clear by now that I regard Bloom's book as a discussion of *Tradition* rather than *Erbe*. He thinks that there are texts we should read not just to understand our cultural heritage but also to abosorb its most significant positive values. Literay scholars have long had a device that seemingly was supposed to do something of this sort. I refer to the reading lists we give our students who are preparing for examinations for advanced degrees. Most discussions of reading lists are not framed in any such terms, to be sure, but Bloom has rendered us a valuable service by making this discussion unavoidable. Before discussing this point further, I offer a limited sample of my emprical evidence.

In reviewing my files, I was able to find four versions of the reading list for candidates for the Master of Arts in German at Brown University, one from ca. 1960, another from 1964, a third from 1969 and finally a revision from 1976. Since about

1980, students make up their own lists, subject to approval by a faculty committee, so the strictly presceiptive list no longer exists at Brown. Appendix A lists authors and/or texts that appeared on all four of these lists. It is not surprising that many younger authors did not appear on earlier lists, but even llowing for this difference, the core selections probably say more about the particular individuals in our deparment and about the oddities of the history of our discipline than about a systematic attempt to define a canon. Lists from other institutions would accordingly differ in many ways.

Names from the earliest list that are abosent from the latest include: Logau, Gerhardt, Gleim, Gellert, Leibniz, Hamann, Leisewitz, Schleiermacher, von Liliencron, Barlach, Benn and LeFort. Names or titles added (apart from the post-1945 category) include Meier Helmbrecht, Wittenweiler's *Ring,* Mechthild von Magdeburg, Meister Eckart, Heinrich Seuse, Redentiner Osterspiel, Reinke de Vos, Rebhuhn's *Sosanna, a chapbook*, Winckelmann, Kant, Gerstenberg, *Die Nachtwachen des Bonaventura*, Raimund, Nestroy, Grabbe, Hoffman von Fallersleben, Freilgrath, Glassbrenner Herwegh, Weerth, Morgenstern, Heinrich Mann and Horváth. In many cases, only selections or a few poems have been added, but the tendency toward addition rather than subtraction is clear.

The 1976 revision of the reading list coincided with the earliest phase of our exchange program, before the signing of an official agreement, but after the beginning of a library exchange and after the introduction of courses on GDR literature at Brown by Duncan Smith. The new section on *Literatur seit 1945* is included here as Appendix B. The list offers 86 names with guidleines for selection of representative authors from each of several categories. Fifty names are from the category "West Germany, Austria and Switzerland" and 36 from the category "East Germnay." This probably was, and may still be a uniquely high representation of GDR authors, and as such it represents a very concrete fruit of our exchange program.

The challenge for American Germanists in the wake of Bloom's book and the discussion that it has provoked is to consider what principles are followed in defining readins required of graduate students training for the professioin. It might be useful to compare how canons are formed in our two societies with different social systems and different understandings of what readin lists are supposed to do. Discussion following the presentation of the oral version of this paper, for example, revealed that the canon for English and American literature in Rostock is determined by negotiation within the faculty, while in German the texts to be read at various levels are established on a national level. It suggests a stimulating topic for a four-way discussion among English and German faculties both at Brown and at Rostock.

Selected Bibliography

Bloom, Allan 1987. The Closing of the American Mind: How higher education has failed democracy and impoverished the souls of today's students. New York: Simon and Schuster.

Colomb, Gregory G. 1989a. "Cultural Literacy and the Theory of Meaning: Or, What Educational Theorists Need To Know about How to Read." New Literary History 20, 457-463.

Colomb, Gregory G. 1989b. "Response" (To Hirsch 1989). New Literary History 20, 457-463.

Hirsch, E.D., Jr. 1987. Cultural Literacy: What every American needs to know. Boston: Houghton-Mifflin.

Hirsch, E.D., Jr. 1989. "From Model to Policy" (Response to Colomb 1989a), New Literary History 20, 451-456.

Hirsch, E.D., Jr., Kett, Joseph F., and Trefil, James (editors) 1988. The Dictionary of Cultural Literacy: What every American needs to know. Boston: Houghton-Mifflin.

Simonson, Risck and Waller, Scott (editors) 1988. The Graywolf Annual Five: Multi-Cultural Literacy. Saint Paul: Graywolf Press.

Swaffer, Janet K. 1989. "Curricular Issues and Language Research: The Shifting Interaction." ADFL Bulletin, 20:3 (April 1989), 54-60.

Zahler, Diane and Zahler, Kathy A. 1988. Test Your Cultural Literacy. New York: ARCO.

APPENDIX A
Reading Lists for the Master of Arts in German at Brown

Authors/texts that occur on all four versions of the list (ca. 1960, 1964, 1969, 1976).

I. Mittelalter und Reformation

Poetry in Echtermeyer/von Weise
Nibelungenlied
Hartmann von Aue, *Iwein* (summary), *Der Arme Heinrich*
Gottfried von Strassburg, *Tristan*
Wolfram von Eschenbach, *Parzival* (summary)
Johannes von Tepl, *Der Ackermann aus Böhmen*
Sebastian Brant, *Das Narrenschiff* (selection)
Martin Luther, *Sendbrief vom Dolmetschen*, Bible translation (selection)
Hans Sachs, one Fastnachtspiel

II. Barock, Aufklärung

Poems in Echtermeyer/vonWeise by selected authors
Grimmelshausen, *Simplicissimus*
Martin Opitz, *Buch von der deutschen Poeterey* (selection)
Andreas Gryphius, one play
Lessin, 17. *Literaturbrief, Minna von Barnhelm, Emilia Galotti, Laokoon* (selection), Nathan der Weise
Herder (excerpts)
Wieland, *Musarion*

III. Sturm und Drang, Klassik, Romantik
One Sturm und Drang Drama (selection from a list)
Poems by Goethe, Schiller, Hölderlin, Brentano
Goethe, *Die Leiden des jungen Werther, Egmont* or *Götz von Berlichingen, Iphigenie auf Taurus, Faust I* and *II, Wilhelm Meisters Lehrjahre, Novelle, Dichtung und Wahrheit IX-XII*
Schiller, *Die Räuber* or *Kabale und Liebe, Maria Stuart, Wallenstein*
Novalis, *Heinrich von Ofterdingen*
Kleist, *Prinz Friedrich von Homburg, Michael Kohlhass*, two further works
Brentano, *Geschichte vom braven Kasperl und dem schönen Annerl*
Arnim und Brentano, *Des Knalben Wunderhorn* (selections)
Tieck, *Der blonde Eckbert*
E.T.A. Hoffman, Two works
A.W. and Friedrich Schlegel, (excerpts)

IV. Zwischen Romantik und Naturalismus
Poems by Droste-Hülshoff, Heine, Lenau, Meyer, Mörike, Platen, Storm
Grillparzer, One play
Heine, *Die Harzreise*
Droste-Hülshof, *Die Judenbuche*
Büchner, *Dantons Tod, Woyzeck*
Stifter, Two works
Hebbel, Two plays
Keller, Two works
Storm, *Der Schimmelreiter*
Meyer, One work
Fontane, One work

V. Die Moderne
Poems by Brecht, George, Heym, Hofmannsthal, Rilke, Trakl, Werfel
Nietzsche, (selections)
Hauptmann, Two works
Wedekind, One work
Schniztler, One work
Kaiser, One work
Musil, *Drei Frauen*
Barlach/Sternheim, Werfel, One work
Kafka, three works
Thomas Mann, Four works
Hoffmannsthal, Two works
Brecht, Two plays

VI. Literatur seit 1945 (Did not exist as a separate category until the 1976 list.)
Borchert, *Draussen vor der Tür*
One play and one novel from a list of authors
Poetry by Bachmann, Celan, Enzensberger, Krolow

APPENDIX B

The Section "Literatur seit 1945" Added to the 1976 Reading List

Group VI. Literatur seit 1945

The literay and ideological diversity of post-war literature in West and East Germany, Austria and Switzerland defies canonization of particular authors and works. You should be familiar with some major literary trends and authors since 1945 and, for the purpose of examination, have an intimate knowledge of at least 3 substantial prose works, 3 plays, and some poetry of your own choice. Your selections should include works from area I and II in each category. The following list of authors may serve as a guide for identifying your interests.

I. West Germany, Austria and Switzerland

1. Prose: A. Andersch, I. Bachmann, P. Bichsel, H. Böll, R.D. Brinkmann, G. Elsner, H. Fichte, M. Frisch, G. Grass, M. v. der Grün, P. Handke, W. Hildesheimer, W. Koeppen S. Lenz, R. Rasp, K. Struck, G. Wallraff, M. Walser, G. Wohmann.
2. Drama: T. Bernhard, W. Borchert, T. Dorst, F. Dürrenmatt, M. Frisch, G. Grass, P. Handke, R. Hochhuth, F. Hochwälder, H. Kipphardt, F.X. Kroetz, M. Sperr, M. Walser, P.Weiss, J. Ziem.
3. Poetry: I. Bachmann, P. Celan, F.J. Degenhardt, F.C. Delius, G. Eich, H.M. Enzensberger, E. Fried, E. Gomringer, G. Grass, H. Heissenbüttel, E. Jandl, K. Krolow, H. Piontek, G. Rühm, P. Rühmkorf, N. Sachs.

II. East Germany1. Prose: J. Becker, J. Bobrowski, V. Braun, G. deBruyn, F.R. Fries, S. Heym, H. Kant, G. Kunert, D. Noll, U. Plenzdorf, B. Reimann, R. Schneider, A. Seghers, E. Strittmatter, A. Wellm, C. Wolf.
2. Drama: P. Hacks, H. Lange, H. Müller, E. Strittmatter.
3. Poetry: E. Arendt, J.R. Becher, W. Biermann, J. Bobrowski, V. Braun, H. Czechowski, F. Fühmann, S. Hermlin, P. Huchel, H. Kahlau, S. Kirsch, G. Kunert, R. Kunze, K. Mickel, G. Steineckert, P. Wiens.

Research on Social Life in the Former GDR

Marilyn Rueschemeyer

My work in the German Democratic Republic during the last fourteen years has focused on the relation of the individual life to the larger economic, social and political structures. Thus I have investigated such issues as the impact of work on family and friendship relations, the attempts of professionals to moderate work obligations at the lowest organizational level, and the response to mass organizations in the new towns. The past project, part of an ongoing study of residential life in the new town, fits in well with a study sociologists at the Wilhelm Pieck University were conducting in the new residential areas around Rostock. Their work, which was coordinated with a number of other studies throughout the country, also provided me with opportunities for contacts in Dresden, Weimar, Berlin and Dessau.

From its very beginning the Brown-Rostock exchange has greatly facilitated my work both in Rostock and in other parts of the GDR. I had access to social scientist, architects and planners, professionals in a wide variety of professional fields, and representatives of political and social organizations. My interviews and visits to a variety of social institutions provided a wealth of important information for my research.

However, one must also note that it was very difficult to do sociological research in the GDR. After the Second World War, sociology in the German Democratic Republic had a different beginning compared to many other fields of study. During the Stalinist phase of the Eastern European state socialist systems, sociology (together with social and cultural anthropology and political science) was considered a bourgeois discipline and a competitor of Marxist-Leninst studies. It was only in the last fifteen years that sociological investigations became more differentiated from Marxism-Leninism. The orientation of research was applied; the focus was on issues of contemporary concern in the country. But it was not the intellectual outlook and the institutional location of GDR sociology that created the greatest difficulties. The most important problems derive from the fact that there is considerable reluctance to make research results freely available. In fact, the results of studies done by GDR social scientists were frequently formally classified as confidential. The researcher, therefore, even when provided with a great deal of material, was well aware that not all internal research had been reviewed and that it was necessary to make assessments without this information.

This paper will offer a sketch of some of the research I have done these past years. It will describe it in the context of the transformations that had taken place in GDR society, transformations that are related to a number of important political and economic developments. The GDR had evolved internally as well as in response to its changing relations to the Federal Republic and to the larger international context. Most recently, the reforms in the Soviet Union and Eastern Europe provided a number of challenges for the GDR, which were interpreted in a variety of ways in different segments of GDR society. These changes in one way or another affected family, work and residential life, the main foci of my research, even before the opening of the border and subsequent unification.

Work and Personal Life: The Transformation of Women's Roles and the Changing Family

Sociologists working on the roles of women generally detail the advances in education and at work that women in the GDR had made; but they remained concerned with ongoing problems that were both culturally embedded and structurally induced, problems that affected the potential for further transformation.

Since the time of the Soviet occupation of East Germany, ordinances encouraging women to become educated and enter the work force helped relieve the desperate need for labor and brought women into the public sphere. At the same time, services and childcare centers were undeveloped; many women organized these themselves in the factories. Beginning in the 60's and becoming more comprehensive in the 70's, social policy addressed a number of issues that were sources of difficulties for families. The new measures included increased development of nurseries and kindergartens, increased leaves and supports after the birth of a child (a year's leave with full pay and a guarantee of being able to return to one's position at work for the first child, and further leaves for each additional child), improved health care, longer work leaves and vacations, modernization and extensive new construction of housing, and an economic policy generally more oriented toward the needs of consumers. The regime hoped that these new policies would increase overall satisfaction, encourage family stability and growth, and improve efficiency and flexibility at the workplace. [1]

These encouragements for women to enter and remain in the workforce and supports outside of work also resulted in a dramatic upgrading of women's educational qualifications. The number of women in university and vocational training was equal to that of men. (Table 1)

[1] Rueschemeyer, Marilyn, "New Family Forms in a State socialist Society: The German Democratic Republic" JOURNAL OF FAMILY ISSUES, Vol. 9 No. 3, Sept. 1988, pp.354-371.

TABLE 1

Percentages of Female Students in Higher Education 1951-1983

Year	All Students	in Residence	Correspondence Students
1951	21.3	23.4	5.2
1960	25.3	31.4	8.2
1970	35.4	43.2	15.9
1975	48.2	56.0	21.4
1980	48.7	52.8	26.5
1983	50.0	53.0	28.5

Source: *STATISTISCHES JAHRBUCH DER DDR 1984* (1984, p. 303).
Note: Not shown: Evening classes

Women who are now forty years and younger had achieved the same educational levels as men. By 1980 already, 70 percent of all women had completed vocational training.[2] To improve the qualifications of older women, since 1973, women over 35 did not have to take a written exam to be classified as a skilled worker if they had already worked at their job successfully for three years. There was also a well-developed extension program for further education that was used to increase one's skills in a particular area, complete a higher level of study, or retrain for another position altogether. The success of this program had been somewhat limited in the last few years because of the difficulty of placing everyone at a

---

[2] STATISTISCHES JAHRBUCH DER DDR, 1981, Berlin: Staatsverlag der Deutschen Demokratischen Republik, pp. 296-297.

job appropriate to the level of qualification; approximately a quarter of the skilled workers and professionals are said to have been working below the level for which they were trained.[3]

Women not only entered the workforce but responded to new opportunities available to them. These were partly the result of structural transformations of the economy as a whole, but they also reflected genuine advances in the relative changes of women. In 1955, about a quarter of all working women were employed in agriculture; that number was reduced to 10 percent by 1980.[4] Before unification, about 30 percent of all women worked in manufacturing industries; and women came to dominate public services (i.e. health, education and social service) as well as trade and sales.[5] They comprised nearly half of the doctors and a third of the legal profession. They entered such traditional male fields as metallurgy and engineering. Altogether, nearly 90 percent of all women eligible to work outside the home did so, about a third on a part-time basis (typically that means they are working six rather than eight hours each day).

These very positive advances were paralleled by remaining difficulties women encountered both in education and at the workplace. To take one example, a recent detailed study documented that the differences in attitude to male and female students remained stronger than one would have expected. As a result, girls were in many cases, still encouraged to engage in

---

[3] Personal conversation with a social scientist.

[4] STATISTISCHES JAHRBUCH DER DDR, 1984, Berlin: Staatsverlag der Deutschen Demokratischen Republik, p.109.

[5] Meyer, Gerd, "Frauen in den Machtierarchien der DDR" DEUTSCHLAND ARCHIV 3, 1986, p.305.

"female" social activities in their free time and discouraged from preparing for "male" occupations in greater numbers.[6]

In the workforce, even with the greater availability of a variety of jobs for women, their positions were concentrated in the lower-paid sectors of the economy (trade and light industry); in heavy industries where the pay was especially good because of "high risk", women were generally excluded. Within each industry, women were segregated vertically in the lower reaches of the occupational hierarchy. This was true for the professions as well. Women were underrepresented at the highest levels of their professions. For example, in university teaching, they were most strongly represented in the lower ranks in the academic hierarchy, and in medicine they were rarely to be found doing surgery or heading their departments. In enterprises, the higher the rank, the lower the representation of women.

Some of the problems women had achieving high rank at work were due to the responsibilities they had in the family, which will be discussed in more detail below. These responsibilities and commitments often meant that women would refuse to accept higher rank if it involved further training, long absences from home, attendance at meetings outside the workplace, etc. Even if they were willing, however, to take these high ranking, prestigious positions, they were sometimes confronted with prejudice on the part of their male colleagues who were reluctant to appoint women to high-level leadership positions. Furthermore, high rank in professional organizations and in enterprises as well as in significant mass organizations

---

[6]Nickel, Hildegard, "Sex Role Socialization in Relationships as a Function of the Division of Labor", QUALITY OF LIFE IN THE GDR: Changes and Developments in a State socialist Society, edited by Marilyn Rueschemeyer and Christine Lemke, New York: M.E. Sharpe 1989, pp.48-58.

provided opportunities and access to high level political positions. It may be that most women (for a wider variety of reasons that were enumerated above) were reluctant to enter political life in the GDR. However, the conditions limiting occupations advancement are also important in understanding the fact that women were under-represented in high political positions in the GDR. One third of all leadership positions in the mass organizations, work and government were held by women; however, the higher the position the lower the percentage of women.[7]

These observations are not meant to denigrate the real and important changes that had taken place; they were, however, a part of the complex transformations that had taken place in the GDR and indicated remaining contradictions in that society.

Marrying at a young age while in school or in vocational training was not at all unusual in the GDR. In 1979, the average age at first marriage was 23.3 for men and 21.3 for women.[8] Access to an apartment of one's own and increasing independence from one's own parents made such a decision an attractive alternative to "single" life. At the same time, men and women who lived together without marrying were a normal phenomenon in GDR life, and this indicates a major change in attitudes. A representative survey in the early 1980s found that 28 percent of the age group 18-40 lived in such relationships. More than half of these couples indicated that they planned to marry later, so that for many of them, this time together is as in the West, a temporary period.[9]

---

[7] Meyer, op. cit., p.306.

[8] STATISTISCHES JAHRBUCH DER DDR 1984, op. cit., p.369.

[9] Gysi, Jutta, "Zur Lebenweise der Familie der DDR", Demographie Seminar 1983, Akademie der Wissenchaften der DDR, p.4.

Approximately a third of all marriages ended in divorce. The reasons for the break-up of the relationship varied enormously as they do in the West, but the consequences were not as severe for the partner remaining with the children (usually the woman). Problems such as loneliness and over-burdening existed, along with the loss of a second income. However, access to work, to an apartment and to childcare facilities as well as other supports enabled the woman to continue on her own. With unification, these supports are endangered.

In my interviews with professional men and women, a number of difficulties which strained the relationships in the family became apparent. These included first the absorption and attention of both partners to their work, which was greater the more skilled and meaningful the job. (There was even more strain in the American families where only the man worked outside the house; indeed dual career couples were far better off than their single career counterparts).[10] However, when there are difficulties at work or other conditions - positive and negative - that carry over into family life, these affect not only the individual involved but also the whole atmosphere at home.

Secondly, the tasks after the long work day - shopping and the care of the children and household - remained primarily the burden of the women. Even though there had been a considerable increase in sharing among the younger and more educated couples, the traditional expectations with respect to gender had not been eliminated. Women, however, tended to expect more cooperation from their husbands than the husbands expected of themselves.

---

[10] Rueschemeyer, Marilyn, PROFESSIONAL WORK AND MARRIAGE: AN EAST-WEST COMPARISON, New York: St. Martin's Press 1981 (paperback 1986).

Thirdly, there had been changes in the self-perception of women which were more significant the more involved women are at work. Women became more confident and independent. Even if they began their marriage with relatively traditional ideas, women increasingly began to question the legitimacy of male dominance at work and at home. It is interesting that while published analyses were steeped in the conditions and perceptions of the GDR, the articulation of dissatisfaction in recent years among professionals writing on the issue of women had also been influenced by the western feminist movement. In turn, western feminists in the Federal Republic have paid close attention to social policy developments in the GDR.

Fourthly, though the state gave verbal support as well as concrete supports for the maintenance of a stable family life there were a number of inconsistent goals and policies which were indeed disruptive. One clear example was the increase in shift work which was viewed by managers and planners as vitally important for increased economic efficiency. However, many of those affected complained that their family life was not working well and that their children as well as their personal relationships suffered.[11] All these factors induced strain and provided some background for understanding new patterns that developed.

---

[11] Dennis, Michael, "Individual Burden - Societal Requirement?" in QUALITY OF LIFE IN THE GDR, op. cit., pp. 95-114

New Residential Communities in Rostock: Issues of Privacy and Social Participation

Looking at pictures of the new housing developments surrounding large East European cities typically results in a somewhat mixed response from the foreign viewer: rather bare and cement-like and at the same time impressively modern. The new town surrounding Rostock had the reputation among GDR residents as being more attractive than most. Furthermore, their location close to Warnemünde, a seaside resort and the main port of the GDR, made them look less than some other settlements as if they were arbitrarily dropped from the air. I was very interested to see how these new communities fulfilled earlier socialist ideals of residential living and how they were architecturally and socially transformed over the years.

Life in these communities was of course, as mentioned above, also a matter of concern within the GDR - to sociologists, to the architects, some of whom lived in the new towns themselves, to the state, which was held responsible for the severe housing shortage and continued to be held accountable for many of the problems that developed in the new communities, and to the residents who comprised 60 percent of the approximately 250,000 residents of Rostock. In the years following construction of the new settlements, there was a great deal of public discussion - formal and informal - of a wide variety of issues revolving around housing and community life that were crucial for at least some of the changes that took place in subsequent developments, for example the size of the apartment houses, the form of the blocks, and the available facilities.

After the end of the Second World War, 35 percent of the urban housing stock had been demolished. However, because of the GDR's own economic priorities as well as the reparations demanded by the Soviet Union, the GDR concentrated its financial resources on construction primarily in the industrial sector. There was some immediate reconstruction of some of the damaged houses, important buildings and streets, but there was little funding available for new housing or for modernizing the existing housing stock. As late as 1981, only 35 percent of the houses were built after 1945; 46 percent of all housing was built before 1919 while 19 percent was constructed between 1919 and 1945. In those houses constructed before 1918, only 5 percent had central heating and only 25 percent had a toilet or bath in the apartment. In 1981, only 25 percent of the housing in the GDR had central heating.[12]

The poor condition of housing was a source of severe dissatisfaction among GDR residents; both the low birth rate and the unwillingness of workers to relocate were attributed to the poor condition of the housing. The long-term housing proposal adopted by the Eighth Party Congress of the *Sozialistische Einheitspartei Deutschlands* (SED) in 1971 prepared both the construction of new housing and the modernizing of old apartments and set the goal of eliminating housing as a social problem by 1990. Through shortages of private living space remained for single adults and young couples, the availability of housing had increased enormously in the last twenty years. Indeed, families who were willing to move to the new communities were delighted for the most part with finally having a comfortable and warm private apartment equipped with central heating and inner bathroom. Rent was subsidized at amounts anywhere from 3 to 7 percent of the family income; it was even lower for those who were unable to afford these rates.

---

[12]Meltzer, M. “Bau und Wohnungswesen” , Simmerman, H. Ulrich H., Felauer, M., ed.s, DDR HANDBUCH, Koln: Verlag Wissenschaft und Politik, 1985, p.156.

Although the socialist way of life in the residential communities was less clearly articulated than the ideas about the organization of the workplace, planners and architects were asked to establish the pre-conditions for a successful family life, for the equal participation of women in the public sphere, for the education and socialization of the young, and for the care of the elderly and disabled. Guided by older models in the Soviet Union and other East European countries and by the advice of such central research institutions as the *Bauakademie* in Berlin and the architecture school in Weimar, each residential community ideally included homes for the aged, clinics, markets, restaurants and buildings for culture and recreation, libraries, nurseries and kindergartens, youth clubs, sports facilities, and other necessary services. Typically, the community - comprising , in the case of the six new towns around Rostock, about 25,000 inhabitants - received the apartments and the most necessary institutions, but was faced with the reality of missing "less crucial" institutions such as shops, a film theater, and other facilities that would make everyday life more pleasant. At the same time, the residential areas around Rostock had improved over the years, and in at least five of the towns, there had been care and attention given to landscaping and surrounding green areas that moderated to some degree the effects of the large concrete apartment buildings.[13]

The main complaint of residents was that there was little to do in the residential area itself. These problems were especially acute for particular social groups. The elderly were

---

[13]For further detail on the architecture, see Derek Bradford and Marilyn Rueschemeyer, "The New Towns Around Rostock," THE JOURNAL OF ARCHITECTURAL AND PLANNING RESEARCH, Spring 1988, pp. 49-63.

more bound to their apartments, often did not have cars, and had difficulty getting to the countryside or to the sea during the weekend, when most families left the residential area. Single parents with children returned from work and had difficulty leaving for the city during the work week because of their children. At the same time, there was limited "entertainment" within the community - cafés, lectures, etc., and there were almost no clubs addressing the needs of single parents. Adolescents had, perhaps, the most severe difficulties. Although there were a few clubs in the residential areas, there were not enough of them. Furthermore, adolescents were not particularly interested in activities that were officially sponsored so that even if the official youth organization, the Free German Youth (FDJ) had become more involved the the residential community, their programs would have had to have been subtle enough and left in the hands of the young people themselves so as not to discourage their participation.

A number of intellectuals and others increasingly interested in more space and variety were attracted to the inner city and to the possibility of having a private house in small towns and the countryside. However, though there was some minor segregation in the residential area - in a few, there was a concentration of people with many children and with some social problems - the typical new town in most GDR cities was very different from the low, middle and upper quality apartment buildings we might have compared them to in the United States, and its social composition was different from that of American suburban areas as well. The new towns in the GDR were much more integrated with respect to social class. Partly due to this more balanced social composition, social problems were not as intense as in public housing projects in the States, although some larger residential areas - such as Marzahn in Berlin with a population of more than 150,000 residents - did have gangs that have engaged in anti-social behavior and even used Nazi rhetoric.

A number of residents had a tiny garden near their house, or in short driving distance, which for some had developed into tiny summer houses.[14] The little garden made living in the new towns more tolerable, improved the fruits and vegetables available to the family and brought in additional income if there was a surplus. During the weekend, people typically spent their free time outside of the residential area, working in the garden, visiting friends and walking. The recent renovation of the inner city of Rostock attracted more people to the cafés and shops in the center of town, though visits to the theater and cinema had been seriously reduced. During the week, television viewing was very popular. Nearly all GDR citizens had access to West German television as well as to the GDR programs.

Within the communities themselves, there were a number of ways to become socially involved. At the origin of the set of official organizations there were varying opinions about how communal these residential areas should be. It is probably fair to say that while the "collective" was strongly developed at work, there was resistance to more collective participation after work in the residential area. Furthermore, it is true that the collective institutions that did exist had only a limited capacity to bring about changes in the residential area; and if they were unable to accomplish much, they had difficulties attracting broad support for their activities.

Just to give a few concrete examples of how this worked, we may begin with the individual building. Families in the new towns as well as in the inner city were organized into

---

[14]For details on private, communal and state housing see Marilyn Rueschemeyer, "New Towns in the German Democratic Republic" in QUALITY OF LIFE IN THE GDR, op. cit., pp.117-143.

Hausgemeinschaften, house organizations. These were to help maintain order, cleanliness and stability in the living area, but also to give support to the political programs introduced and to generally encourage a "socialist way of life." The ideal house organization comprised not more than twelve families and contained a mixture of apartment sizes to encourage mutual help among people of different ages.[15] In reality, many *Hausgemeinschaften* were much larger. The success of these *Hausgemeinschaften* varied enormously. When people came from very different social backgrounds, when they did not adhere to the norms of the apartment house, when there was a lack of leadership, or when the house-community was not well-defined, the residents came together primarily to decide on necessary repairs or to take care of an emergency that may arise in the apartment house. Furthermore, the *Hausgemeinschaft*'s requests for funding or for repairs might be ignored for long periods of time. When responses to complaints from the residents were not satisfactory, when repairs were not funded, people became discouraged and less active in the *Hausgemeinschaft*. Certainly in most buildings not everyone responded to campaigns to clean up or repair, or properly care for the grounds.

The *Hausgemeinschaft* rarely had the flavor of the successful work collective in the enterprise. At the same time, with few exceptions, both the inside and the outside of the buildings, at least in the houses I saw in Rostock, were in moderately good condition.

The Democratic Women's Federation (DFD) was an example of mass organization in the new towns that participates

---

[15]Staufenbiel, Fred et al., SOZIALOGISCHE UNTERSCHUNG VON WOHNGEBIETEND DER STADT ROSTOCK, Weimar: Hochschule für Architektur und Bauwesen 1983, p.130.

in traditional education activities of various sorts and in addition joined with other groups to aid the elderly and others in need. It also sponsored lectures of different kinds. This organization, even with its accomplishments, had difficulty attracting large numbers of women. Aside from it projects with the elderly and its educational functions, the DFD seldom took on activities that were of real benefit to the women themselves, and it rarely challenged official policy and the official interpretations of women's issues. This group had more success when it addressed issues residents were immediately concerned about such as adequate public transportation, scheduling of shop openings, etc.. Similar difficulties faced the National Front, an umbrella organization which included the leading Communist Party, other subordinate parties as well as mass organizations and social groups.[16]

The fairly low level of participation in these organizations does not reflect the quality of social life and the degree of interaction that took place in the residential areas. Social life in the residential community was primarily informal in nature. It involved help, aid, and friendship in a wide variety of situations. Organized and mandatory participation was avoided and even shunned in the residential community. In contrast to the workplace, the world of residence was seen and protected as the place of privacy, family and friendship, and voluntary informal relations.

Family life, institutions of work as well as housing and residential life continued to evolve and change in the GDR. These changes came about in part because of how people

[16]For further detail on these issues, see Marilyn Rueschemeyer, "Participation and Control in a State Socialist Society: The GDR." Center for European Studies, Harvard University, Working Paper No. 17, 1989; revised version in East/Central Europe, forthcoming.

perceived their needs and their situation and how they acted in their everyday lives. They were also affected by the policies of state and party, by economic change, and by ways in which the different official organizations responded to the voiced needs of people. These complex interrelationships were continuously changing. They provided a fascinating arena for research on personal life in a changing socialist society. The Brown-Rostock exchange program gave social scientists the opportunity to exchange views on these issues and to become part of an ongoing research conversation. It enabled both sides of the exchange to move beyond the stereotypes we held of each other. This process itself has had a profound effect on all of us.

Library Exchanges - a Contribution to the Promotion of International Understanding

Karl-Heinz Jügelt

Nowadays, our library has more than 1,800 exchange partners in over 80 countries, 50 of which are in the U.S. Some of these date back to the middle of the last century.

Our first information about exchange of publications itself dates from the nineteen-thirties. We know that national exchange of dissertations and other university publications was established in Germany in 1817 as Academy Exchange Association (*Akademischer Tauschverein*) and we also know that the University of Rostock was involved. At this time, the exchange of university publications was handled by the office of the rector. Later on the exchange became the duty of the University Library. According to some of the materials found in the archive of our University Library, the first exchange with an American institution was established in 1853 when the Smithsonian Institution in Washington D.C. sent some books, and in 1875, the Department of War of the North-American Free States, in a letter dated February 2, asked for Rostock medical dissertations. The numerous publications from American universities from the 19th century in our collections gives us an impression about the many exchange relations between Rostock and universities in the United States in the past.

After the Second World War and especially during the last two decades in many of the fields of science and technology, international cooperation has received fresh impetus. This is also true of library science in general and the exchange of publications in particular.

This increase derives from the convention concerning the international exchange of publications and the convention concerning the exchange of official publications and government documents between states of UNESCO (1958).

Libraries are now faced with the rising demand for oder out-of-print and for new non-trade publications as well as for foreign books for which foreign currency may be scarce or unavailable. All these require greater efforts on the part of the exchange librarian. Nowadays it is extremely gratifying to observe how the technique of exchange is proving to be of increasing value and importance in our University Library acquisitions efforts. Not only does it continue to fulfill with undiminshed vigor its traditional role as a ssupplement to regular purchase activities, but also it is being looked to as a steady resource for acquisitions when the University Library purchase budget is being threatened by restrictions based on a shortage of foreign currency. In the face of these added demands, it is all the more remarkable that the University Library has been able to maintain its exchange program as a healthy and viable component of its total acquisitions picture.

In the fiscal year 1988, the University Library received 41.7 percent of the new acquisitions through world wide international exchange and 1988 was a typical operational year for University LIbrary's exchange program.

Having said all this by way of prologue, we have to look into the legal basis of exchange programs for GDR libraries. Although it is true that libraries have been exchanging materials for more than 150 years, the legal basis for these activities was finalized, after a long and thorough preparatory stage, in the "regulations for the international exchange of publications of libraries and information centers and the exchange of publications of libraries and information centers and the exchange and delivery of official publications and governments"

-published in the *Gesetzblatt der Deutschen Demokratischen Republik* , part I, no. 12 of 12, April 1976.

By these regulations, the following advantages are offered for the development of exchanges:

1. A long term orientation for the exchange of publications as a part of applied foreign policy

2. A maximal standardization of customs regulations for the libraries.

3. Precise definitions of the objects of exchange and a legally binding definition of the term "official publications"

4. The legal fixation of the exchange of book trade publications - what is actually exchanged - as we will see later on.

In the preamble of the regulation, reference is made to UNESCO in the exchange of publications and the endeavors of the GDR to contribute to this kind of international cooperation, and the two conventions of 1958 are expressly mentioned.

In Article 1, it is stated that libraries and information centers come within the purview of this decree.

Article 2 defines the international exchange of publications as relations between the institutions of the GDR and other states and it says that the exchange shall be based on the acquisition policy of each individual library.

Article 3 deals with the objects of exchange. In accordance with Artice 2 of the UNESCO conventions, they are: all publications printed or similarly duplicated including doctoral theses, journals and newspapers; moreover, independant maps, art prints, pictures with or without text, printed music, sound carriers, as well as their copies or microforms and official publications obtainable either through or outside of the book trade. The libraries are authorized to exchange their own publications and those of the bodies to which they are attached, the products of the publishing houses of the GDR, literature either sorted out or not yet processed but officially approved, and also literature expressly acquired and books falling within the range of the regulations for the protection of cultural goods are excluded from exchange.

Article 4 specifies the institutions designated to engage in these exchanges. Among them are all important research libraries, such as university and college libraries, central special research libraries, the central libraries of the Academies, the central scientific district libraries, etc. In this connection, the German State Library has been entrusted with the administration of exchanges with foreign state and national libraries within the framework of intergovernmental agreements. Moreover, it is important to keep in mind the libraries attached to university sections or schools and institutions shall handle their exchanges through their central libraries thus ensuring uniformity.

In Article 5 the exchange partners are defined as follows: libraries, scientific institutions, archives, museums, enterprises and public organizations, not, however, private persons, publishing houses or enterprises in the book trade.

Articles 6 and 7 regulate the exchange of official materials in detail.

Article 8 provides the exemption from duty as well as export and import permits, and it also imposes special obligations on the libraries.

In Article 9, the head librarians are obligated to lay down special regulations for the administration of exchanges in their institutions. They should be prepared following a basic guideline issued by the Ministries of Culture and of Higher Education, operative since January 1, 1981.

The libraries must also furnish exact evidence of their exchange statistics to the national exchange center of the German

State Library which, in turn, submits the complete statistics to UNESCO.

On the basis of the Library Regulation of the GDR, dated May 31, 1868, the university libraries collect, process and provide scholarly literature for teaching and research , as well as for the education and training of students. They are also important resource centers within the regional and transregional areas and contribute to the development of research collections within the framework of the "Subject Specialization Plan of the Research Libraries of the GDR."

Whereas university libraries as a rule cooperate in the library networks of the GDR through departmental libraries according to their character as special research libraries, in the framework of the Subject Specialization Plan the responsibility for a special field depends also on the tasks of the institution of the university to which they are attached.

Let me use as an example the responsibilities of our library within the framework of the Subject Specialization Plan. Since only at the university in Rostock is there a School of Marine Technology and and Engineering College of Shipbuilding Technology, and since our School of Biology is especially engaged in teaching and research in the field of Marine Biology, our library is responsible for these fields.

A greater responsibility of our library stems from the fact that Rostock is host to the only School of Latin American Studies in the GDR. Therefore our collection development for teaching and research on Latin American history, economy, social life and social movements, linguistics and literature, music and arts requires not only large amounts of foreign currency, but also a very good exchange of publications with libraries in Latin American countries and with libraries of institutions all over the world dealing with Latin American studies.

As a counterpart we offer to our exchange partners the Academic Journal - published in two parts ten times a year - composed of 20 periodicals of the schools of our university in several fields such as agriculture, biology, library science, philosophy, history, physics and so on.

Our exchange with Brown is a special one. Initiated by Professor Hans-Joachim Bernhard and Professor Duncan Smith at the beginning of the seventies, an official exchange agreement was established in March, 1977.

The monetary value of the exchange was, at this time, $500.00 (at this time about 1030 Marks). At this time, we could establish this exchange with books from the publishing houses of the GDR because according to the legal regulations for exchange of publications from 1976 - as I've mentioned before - our library was authorized to exchange these materials.

One year after this promising start, a review of recent material exchanged between the two institutions indicated the need for changes and Professor Bernhard of Rostock, Professors Schmitt and Smith of Brown, Dr. Dorn of Otto Harrassowitz Buchhandlung Wiesbaden and members of the Brown University staff met in March, 1978. Considering some of the difficulties in the previous arrangement, it was agreed that the exchange would be limited to material in the field of German literature: this means that this agreement was expressly contracted to support research of the Bernhard research team. Therefore it was agreed also that the selection of material exchanged from Rostock to Brown is to be made by Professor Bernhard or his designee. This material should include all primary works of GDR literature (*Belles Lettres*), original publications, revised editions and anthologies should be sent; occasional important reprints (primarily *Edition Leipzig*) may be included only after prior approval by Brown University. All critical literature is excluded except for volumes appearing in the following series:

Edition Neue Texte (Aufbau-Verlag)
Volumes related to German Studies only
Germanistische Studien
Literature und Gesellschaft

Materials exchanged from Brown to Rostock should be works published in German speaking countries. Thus it was possible to establish a strong collection of *Belles Lettres* of the GDR at Brown - in my opinion it may be third after the Library of Congress and Harvard - and a strong collection of *Belles Lettres* of the Federal Republic of Germany at the departmental library of Wilehlm Pieck University, which was then serving as the basis for researching, writing and publishing a very important volume, "History of *Belles Lettres* of the Federal Republic of Germany", edited and to a large degree written by Hans-Joachim Bernhard.

I think this is a very good example of direct support for a researh project given through an exchange agreement.

In the following years, our exchange agreement has been regularly overseen and developed. At first, during my visit to Brown in 1981, I discussed problems of the exchange with Fred Lynden and Janet Hampton and we later proposed a change from fiscal year to calendar year, and to strengthen the GDR collection at Brown, we tried to purchase antiquarian materials for Brown to fill some gaps in the excellent collection.
-Neue Deutsche Literatur (2 ex)
-Latein Amerika

1975	11 Monographien
1977	123 Monographien
1978	31 Monographien
1979	34 Monographien
1980	116 Monographien
1981	65 Monographien
1982	81 Monographien
1983	70 Monographien
1984	185 Monographien
1985	113 Monographien
1986	116 Monographien
1987	74 Monographien
1988	41 Monographien
1989	65 Monographien
	1,123 Monographien

The result - in my opinion - is very good; since 1977 we have sent more than 1,100 books to Brown, which means a yearly average of 80 books. The next opportunities to discuss actual exchange problems were the visits of Fred Lynden in June, 1983 and in September 1985. We discussed problems such as the currency exchange rate, but it seems to me that more important were our discussions about enlarging our exchange program to the fields of sciences mentioned in the contract regarding academic cooperation of both universities (German literature, English and American literature, physics and medicine). According to the developments of this contract, we will need to change our exchange agreement too. Materials going to Brown should include most of the publications of Rostock Wilhelm Pieck University (we presently send only *Lateinamerika*) and Rostock University Library and publications of the GDR book production in the above mentioned fields of science. This may be needed for preparation of studies to be done in Rostock and for full information abour study and research in Rostock.

A new project of cooperation between both our libraries is common exhibitions of rare materials. Although we generally agreed to arrange such exhibitions, there are some difficulties with transportation, insurance and so on. Because we have to consider the status of the materials as an extremely valuable part of our cultural heritage, we also have to consider legal aspects of transportation and appropriate insurance. The types of materials that may be included in the exhibition may be "History of Science", "Treasures of the Two Libraries" or Yiddish manuscripts, or rare books.

I believe that an exchange of exhibitions would be a remarkable contribution by our libraries to the cementing of the Rostock-Brown connections and I hope we will have time enough to discuss all related problems and come to some concrete decisions.

Thus our increasing personal contacts, our exchange of publications and exhibitions will be a remarkable contribution to better understanding and peaceful cooperation between our libraries, our staff and our universities.- as we will see later on.

Libraries and International Exchanges

Frederick Lynden

In 1989 the libraries of Wilhelm Pieck and Brown Universities can celebrate 17 years of working together. As early as 1971/1972 Wilhelm Pieck University shipped novels by GDR authors, and in return Brown sent novels by FRG authors and its journal, Novel, a forum of Fiction. Since that time, there has been a continuous stream of materials, requests and correspondence between the towns on the Seekonk and the Warnow. The library exchange has been further strengthened by visits from Dr. Karl-Heinz Jügelt, University Librarian, Wilhelm Pieck, to Brown in 1981, and Frederick Lynden, Assistant University Librarian for Technical Services, Brown, to Wilhelm Pieck in 1983 and 1985. Today I am pleased that Merrily Taylor, University Librarian, Brown, is present to commemorate this occasion. This library exchange has become more than an opportunity to interchange books and journals. It has also developed into an opportunity to exchange ideas about librarianship. As part of the exchange, Brown receives *Zentralblatt für Bibliothekswesen* and Wilhelm Pieck is sent RLG (Research Library Group) News. Thus, the exchange has meant more than a mere receipt of materials. It has provided librarians with a vital connection to their colleagues overseas.

The library exchange with Wilhelm Pieck has been very important to Brown. Today I will comment on how it has affected the library collections and research at Brown. You should first be aware that the exchange encompasses almost exclusively literature. It is no accident that literature was chosen as the primary focus of the exchange. At Brown, the library exchange originated through the German Department assisted by Duncan Smith and William Crossgrove.

Why is literature the principal focus of the exchange with Wilhelm Pieck? As Ezra Pound noted "Literature is language charged with meaning." I think this conveys the importance of literature to the German department, but it also explains its importance to research. It is often stated in the library that sciences departments have their laboratories for research whereas the library is the laboratory for the Humanities and Social Sciences and primary literature is an important documentation of a society's culture. Indeed it reflects the society in its language, its setting, and its story. Thus, the library considers it very important to collect the literature of the GDR since literature is the tool by which students and scholars can gain a better understanding of the society and culture.

Since the exchange has existed for over 17 years, it is interesting to note that some of the works of the same authors were received in the 1971/1972 and 1987/1988 shipments: Franz Fühmann, Juri Brezan, Wolfgang Joho, Johannes

Bobrowski, Erik Neutsch and Erwin Strittmatter. These writers are clearly established authors of the GDR. Altogether during the first five years of the exchange, Brown received 249 titles. By any measure, the Brown Library has a very strong collection of GDR literature. A recent check of Erwin Strittmatter in the Online Computer Library Center (OCLC) database, showed 24 unique records in this US database which contains 18 million holdings of U.S. Canadian and International libraries. Of these 24 Strittmatter records, Brown holds all but 4 and two of the four are children's books. Therefore, Brown holds approximately 90% of the Strittmatter holdings in the libraries of the U.S. This is but one example of the strong holdings of GDR literature at Brown.

The Brown Library of course, would like to expand its holdings of GDR literature. In 1984, the library increased the exchange dollar amount to $1,000 and sent a list of titles which had not been received since the exchange began in 1972. There were a total of 199 titles on the list and, as of today, Wilhelm-Pieck University has supplied 106 of these titles. The response to the original list was excellent and contained ninety-three titles from the formative period of GDR literature (1970 or before). One of the two titles mentioned above of Strittmatter, which we are still missing, *Damals auf der Farm und andere Geschichten*, is on the Brown desiderata list and we are still awaiting its supply. Some of the items are undoubtedly very scarce such as Christa Wolf's *Moskauer Novellen* and may be impossible to find. Nevertheless, the Rockefeller Library at Brown will own some rare GDR literature titles in time.

Before discussing some of the research results which this collection of literature has made possible, it would be useful to discuss how the exchange operates. At Brown we have one staff member, Mira Eides, who is responsible for exchanges. We currently have exchanges with 254 libraries in 37 countries. Of these 254, four are exchanges with libraries in the GDR, the Wilhelm Pieck exchange being the largest and most comprehensive of these. The exchange specialist is responsible for keeping track of what is received and shipped, claiming materials, and monitoring the costs of the program. For Brown's part, i.e. shipping FRG titles, the Library relies on a FRG bookdealer, Otto Harrassowitz, to supply all FRG literature titles to Wilhelm Pieck University. Very recently the Library has received request for some English language titles, and the Library has obtained these through its customary sources. One of the mainstays of all of Brown's exchanges is the Novel: a Forum on Fiction, a journal of history and criticism published at Brown since 1967 and nationally known as an important critical journal. The Library also sends this journal to Wilhelm-Pieck University.

Over the years, there has had to be some fine-tuning of the exchange. At times, unwanted materials, e.g. children's literature, slip in and have to be returned. Although children's literature has a significance in reflecting the values and morés of a society, the Library at Brown does not collect children's literature. Other issues have been:

1) should the fiscal or calendar year be used for measuring costs?

2) should marks or dollars be used to determine value of shipments?

3) Is the selection at Wilhelm Pieck University of GDR fiction appropriate?

4) Should the limit on total dollars/marks spent be raised?

5) Are the faculty sufficiently satisfied with the exchange?

Some of the issues are easy to resolve. The libraries use the calendar year for the exchange and both measure value in marks and dollars. The last three issues relating to selection, total cost, and faculty participation require some discussion.

First, the issue of selection is one of great interest at Brown. The unofficial exchange began in 1972. In 1978 an official agreement was drawn up after meetings among Professor Duncan Smith, Professor Albert Schmitt, Fred Lynden, Janet Hampton, Dr. Knut Dorn of Otto Harrassowitz and Professor Hans-Joachim Bernhard of Wilhelm Pieck University. Stuart Sherman, Acting University Librarian, signed the agreement for Brown and Dr. Karl-Heinz Jügelt signed for Wilhelm-Pieck. The agreement specified that selection was "to be made by Professor Hans-Joachim Bernhard or his designee and material should include all primary works of GDR literature (*Belles Lettres*); original publications, revised editions, and anthologies should be sent; occasional important reprints (primarily "*Edition Leipzig*") may be included only after prior approval from Brown." Critical literature is excluded with the exception of some series. In checking various sources such as the GDR Bulletin, a newsletter for literature and culture in the German Democratic Republic, published at Washington University in St. Louis, recommendations from the German Department selector and the Library selector, and slips representing titles supplied to Boston Public Library by Harrassowitz. The Library now finds that it is not receiving all primary works of GDR literature of value to an academic library. The problem appears to be twofold: first, it is always difficult to select literature at the outset. That is to say, who can be certain a novel will become significant when it is first published? Many library purchasing plans, where books are received automatically, depend upon a list of established authors assuming that these authors will continue to write works of importance. Naturally, it is always difficult to predict who will become noteworthy so a library is bound to miss significant works. Second, academic libraries prefer to collect as comprehensively as possible, but wish to exclude certain genres of literature such as light romances, mysteries, travel tales, and adventure stories.

According to UNESCO Statistical Yearbook , in 1986 the GDR published 764 titles in literature. During a year, Brown receives on an average 85 GDR literature titles as part of the exchange. It is difficult to estimate how many of the 764 titles an academic library should purchase. Most of the 764 titles, such as light fiction, children's fiction, and fiction of minor authors, are not suitable for an academic library. Otto Harrassowitz, a FRG vendor which supplies books to U.S. academic libraries, supplied only 67 GDR literature titles in 1988 to American academic libraries. Harrassowitz relies on a list of established authors to select such titles for academic libraries. However, despite careful selection, libraries will miss adding titles from significant new authors especially since many

GDR literary works are produced in short runs. These titles must be sought long after they have been published. The library exchange provides an opportunity to obtain these scarce titles.

A second issue is that of cost. In order to provide for searching of scarce titles, it is the time to raise the limit from $500 per year to $1000. However, many may not be aware that the rising costs of serials, particularly foreign serials which are affected by the weakening dollar, has caused the Brown Library to buy fewer and fewer monographs. This situation has, of course affected our exchange programs. The Library and the German Department need to work together for increased acquisitions.

This recommended increase in library exchange funds relates to the third issue--that of faculty satisfaction with the exchange. At the moment, there has been no great dissatisfaction with the library exchange. Means of obtaining necessary materials through inter-library loan and personal purchase have undoubtedly contributed to this agreeable mood. However, the Library believes that a program of determining what is lacking in the collection should be undertaken. The base collection is too strong to allow the entire collection to weaken. This probably means some work on the part of the faculty, the Library, and graduate students and ultimately some help from the Library at Wilhelm Pieck. Ideally the faculty in consultation with library collection development staff should recommend authors the Library should collect. The Library can then search titles and make up desiderata lists. Finally, the Library will use the exchange to obtain the needed titles. Thought should also be given to improving the current selection techniques.

While the exchange needs some improvements, it has provided the necessary materials for a book, an article, two dissertations, and two student honors theses over the past twelve years. The dissertations focus on the work of Volker Braun and Peter Hacks and were done in 1977 and 1979 respectively. The first by Carol Sue Dexheimer is entitled, Peter Hacks' plays: Socialist Comedy in the German Democratic Republic. Her dissertation examines Hacks' plays in the ideological context in which they were written. " He seeks to stimulate GDR audiences to depart happily from a status quo which is anachronistic from the perspective of his Marxist human ideal." By 1977 Brown had an excellent collection of Peter Hacks and we continue to receive materials by Hacks. In 1979, Mary Angela Harrington von der Lippe wrote her dissertation: " The Emergence of the Socialist Individual in the works of Volker Braun". Her thesis "examines the notion of the "socialist individual as it is developed and defined through the functions of the artist, the poet, the worker and the female in Volker Braun's work." Three of Braun's works noted in her dissertation were received from Wilhelm Pieck in 1975. Clearly the GDR literature collection at Brown University was a significant resource in the creation of these two dissertations by Brown graduates.

The collection has also been of significant assistance to one student who has produced a senior honors thesis and one student who is working on her honors thesis. Joseph Polman did his senior honors thesis comparing the treatment of the individual in American and GDR fiction. Polman compared the individual in Ken Kesey's book One Flew over the Cuckoo's Nest with Strittmatter's Ole Bienkopp. He also looked at other GDR and American texts and concluded that "these texts coming out of different languages and cultures reflect a kind of common human experience of individualism." He also perceived the differences in the literary works from both countries "on the US side to focus on the importance of the individual and a tendency on the GDR side to focus on the society." He further discussed the different perceptions of freedom and the work place in the two societies as reflected in the novels. Polman's work was clearly facilitated by the extensive literary collections in the Brown Library.

Another student who is studying GDR literature at Brown is Gillian Silverman. She is focussing on Christa Wolf's Kassandra writing her senior honor's thesis for the department of Modern Culture and Media Studies. In particular, she is considering the literary treatment of Cassandra, heretofore a figure only examined by men. Christa Wolf's novel looks at the roots of warfare through the Trojan war and the Cassandra myth in the context of World War II and Hitler. Gillian Silverman is analyzing the issues of what constitutes mythology and the reappropriation of mythology by Feminists in relation to Kassandra.

There is annually a course offered on the literature of the German Democratic Republic. For the last few years it has been taught by Professor Carol Poore who features Wolf Biermann, Volker Braun, Bertolt Brecht, Christopher Hein, Sara Kirsch, Heiner Müller, and Ulrich Plenzdorf. She covers such topics as literature and the construction of socialism, the avant-garde in life and art, youth under Socialism, and women and men - reality and utopia. This course is also taught by visiting professors from Wilhelm Pieck University. This year, for example, it is taught by Sigurd Schmidt. Without the extensive collection of GDR literature in the Brown University Library it would not be possible to teach such a course. The authors noted above are well represented in the collections. As part of Carol Poore's course there is always a paper to be written in either

German or English which will "include some literary works not covered in the class." This, of course, requires the students to use the collections built through the library exchange between Brown and Wilhelm Pieck Universities. Professor Poore's research on illness in GDR society published in the proceedings of the Eleventh New Hampshire Symposium on the German Democratic Republic also relates to GDR literature. It is entitled: "Illness and the Socialist Personality: Philosophical Debates and Literary Images in the GDR."

Finally, in the case of the book about the GDR, Walls and Mirrors: Western Representations of really existing German Socialism in the German Democratic Republic by Professor Duncan Smith, it is hard to say whether the collection has made the book possible or its author has made the collection possible. Suffice it to say that his book makes much use of GDR literary works and considers their influence on the culture and society. One entire chapter focuses on the literature of the GDR and speaks about such authors as Erwin Strittmatter, Elke Erb, Volker Braun, Günter DeBruyn, Jurek Becker, Stephan Heym, and Steffen Mensching. Duncan Smith was largely responsible for insuring that the library exchange was begun and has helped nourish it through the years with his comments, suggestions, and criticism.

What about the future of the library exchange? It is clear that it will have to be expanded in a financial sense if Brown is to continue to collect GDR literature, since the cost of titles has risen as the dollar has weakened and the Library is really receiving less material for the same expenditure. It would also be useful to continue to track what has been published versus what has been received at Brown in order to fill in lacunae which may have developed over the years. On the other hand, as I have noted earlier, our exchange partnership has proved to be a very effective means for obtaining scarce materials. The percentage return (53%) from Wilhelm Pieck Library on the desiderata list was quite gratifying. It means that there is still a recourse if we do not obtain the appropriate books at the time they are published. There is no question that lasting literature is hard to select at the time of publication unless the author's previous works are well known. On the other hand, it is best to cast as wide a net as possible in order to get the best fiction, the first time it appears. If one gets chaff among the wheat, it can be discarded.

The work of evaluating the collection will be much easier in the future thanks to new library technology. For example, it was possible for me to check the 199 titles on the desiderata list using the Library's on line catalog, JOSIAH, without leaving my office. I did not have to go to the card catalog to individually search the authors on the list. By typing the author's name into my Macintosh personal computer the catalog software searched the author's names in the approximately 650,000 records currently in the file. The searching of the list took less than an hour to do. The PT (or German literature) classification books have not yet all been entered into the computer file. Once all titles by an author held at Brown are entered into the computer, it will be possible to print-off a list of all titles held at Brown of a particular author. This will be ideal for checking Brown's holdings against bibliographies to see the extent of Brown's holdings.

There is another exciting technological device for searching information about GDR literature in periodical articles, dissertations and articles in books. Recently the Library at Brown installed a personal computer with a CD-

ROM device (Compact Disk-Read Only Memory) containing the MLA (Modern Language Association) Bibliography. The Modern Language Association Bibliography file covers the period from 1981 to the present (September 30, 1988) and it includes 450 citations for GDR literature. Among the citations, for example, is the article by Carol Poore. The five inch disk contains a quarter of million MLA records, among them much useful critical writing about GDR literature. The various *Wissenchaftliche Reihen* from universities, such as that of Wilhelm Pieck University, are indexed by MLA. More than 3,000 periodicals and series are indexed.

This paper has covered principally the topic of exchanging books. One of the early suggestions made was to do an exhibition. This planning was further advanced as part of my visits in 1983 & 1985. Since that time, we have exchanged lists of possible titles to be included, and Merrily Taylor and I are here to make further arrangements. We hope to exhibit books from the early history of Science and/or treasures from each Library. The Exhibit would be held in 1990 in both Rostock and Providence and would be a very effective means of publicizing the library exchange.

In summary, the Library Exchange between Wilhelm Pieck and Brown has provided much more than an exchange of literature between the U.S. and GDR. It has offered opportunities to exchange views on research librarianship and will make possible a significant exhibit of rare special collections thus further publicizing the unique arrangement between these two universities. I recall very well the comment of one of my GDR colleagues as I left Leipzig in 1983. As he shook my hand, he said: " *Bibliothekswissenschaft lebt*!" As colleagues we have a common language, realize similar successes, and understand each other's problems. This mutual understanding is the most important component of the Library Exchange. We both are here to provide our universities with the materials for research and thanks to the exchange we are doing a better job.

Research into Thermophysical Properties - a Challenge to International Cooperation: The Brown - WPU Connection

Part I: Joseph Kestin

My cooperation with the team at Wilhelm Pieck Universität followed soon after Professors Ulbricht and Kristen had spent several weeks in Providence, becoming acquainted with their partners at Brown University. I was very glad to accept their invitation to visit Rostock because I knew of the existence of the oscillating-disk viscometer, designed and built by Dr. E. Vogel. A simplified sketch of this instrument is reproduced in Figure 1. My own group, as well as the team in

Rostock had at that time worked on a program of absolute measurements of the viscosity of gases, striving to secure maximum reliability and accuracy. The heart of the instrument is the highly accurately polished disk suspended between two plates and performing torsional oscillations. The very attractive feature to me was the fact that the whole instrument was made of quartz. This had the advantage that it practically eliminated effects of thermal expansion and promised great precision.

On the one hand I was pleased to discover a group of research workers working on an identical program and sharing our ambitions for accuracy, all the more so that they had discovered, and made use of the absolute theory of the instrument developed by G.F. Newell, then at Brown, in cooperation with me. On the other hand, I wanted to build a replica of the Rostock instrument which I could do only with the active cooperation of E. Vogel. I did invite him for an extended stay at brown University and was very disappointed when I was informed that his absence from Rostock could not be arranged in view of the heavy teaching load experienced at the time.

Our cooperation continued by correspondence and exchange of publications. Both groups extended their programs in that they undertook to produce very precise and reliable measurements of the thermal conductivity of gases. Again, an absolute theory of the instrument, based in the hot-wire principle, was worked out at Brown University with the very active cooperation of Dr. W. A. Wakeham, now head of the Department of Chemical Engineering at Imperial College in London. Both groups further enlarged their programs by the inclusion of theoretical studies of the properties of the forces which act between molecules. This could be done very effectively, because the experimental data created in both places were suitable as an aid in the development of theory thanks to their accuracy, reliability and absolute character.

The synthesis between measurement and theory has led to the formulation of what has later become known as the KRW (Kestin-Ro-Wakeham) law of corresponding states.

While this work was in progress it became possible to invite Dr. J. Millat to join the effort. He was able to spend a period of time in Providence, and later with Professor Wakeham who had returned to his native England and joined the faculty of Imperial College of Science, Technology and Medicine in London.

Professor Wakeham developed the Thermophysical Properties Data Center in London in cooperation with the International Union of Pure and Applied Chemistry. In parallel, the writer proposed the formation of a Subcommittee on transport Properties of Commission I.2 of the International Union. The writer became its Chairman, Professor Wakeham was elected Secretary , and Dr. Millat was coopted as Corresponding Member.

With these steps, the cooperative program between Brown University and the Wilhelm Pieck University has become incorporated in a truly international effort. This is testified to by the fact that the IUPAC Subcommittee unites members from thirteen countries: the Federal Republic of Germany, France, the German Democratic Republic, Greece, Iran, Japan, Mexico, The Netherlands, Norway, Portugal, Singapore, the United Kingdom, and the United States of America. A candidate from the Soviet Union has been invited to join the next meeting of the Subcommittee, scheduled for August 11-13, 1989, in Lund in Sweden. The Subcommittee meets regularly once a year for the purpose of exchanging ideas, new measurements, and the formulation of joint projects.

Currently, a team consisting of G.A. Olchowy and Jan Sengers of the USA, of V. Vesovic, W.A. Wakeham and J.T.R. Watson of the United Kingdom together with J. Millat of the German Democratic Republic are putting the final touches on a definitive formulation of the properties of carbon dioxide (the "green-house" gas) under the auspices of the International Union.

The IUPAC Subcommittee on Transport Properties will sympathetically consider an invitation to hold its 1990 or 1992 meeting in Rostock.

The almost ten years of active cooperation resulted in a number of joint publications whose theoretical and practical importance will be discussed in Part II.

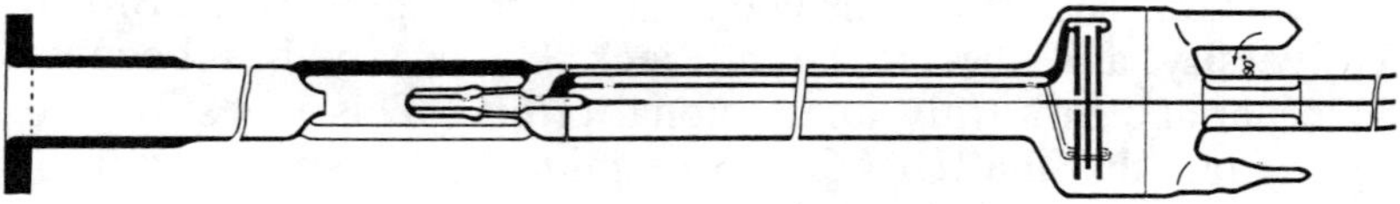

Figure 1. The oscillating-disk viscometer, made wholly of quartz, designed and built by E. Vogel of W.P.U. (Reprinted from Wissenschaftliche Zeitschr. der Universität Rostock, 1972)

"Vom Nutzen des Dialogs und der Wissenschaftlichen Zusammenarbeit - 10 Jahre Austauschprogramm Wilhelm-Pieck-Universität Rostock/ Brown University Providence"

Part II: Jürgen Millat

Es freut mich außerordentlich, daß ich zu diesem Seminar beitragen kann. Das ganz besonders auch deshalb, weil mein Beitrag ganz nahtlos an den eben von Professors Kestin gehaltenen anschließt, gleichermaßen demonstriert, daß wir nicht nur über Kooperation reden, sondern daß sie schon soweit Alltag geworden ist, daß wir auch ohne vorherige große Abstimmung auf Tagungen auftreten können. Dieser Sachverhalt, das sei vorweggenommen, gilt übrigens nicht nur für diese bemerkenswerte Zusammenkunft, sondern auch und vor allem für unsere Mitarbeit bei internationalen Tagungen, immer mit dem Ziel, gemeinsam einen Beitrag zum Fortschreiten unserer Wissenschaft zu leisten. Ich komme auf diesen Punkt noch zurück. Natürlich erstreckt sich unsere Zusammenarbeit auf ein speziales Gebiet, es ist im Titel unseres gemeinsamen Beitrages genannt, ein sehr vielschichtiges Gebiet, das großes Anteile an naturwissenschaftlicher Grundlagenforschung mit wichtigen Anwendungsaspekten verbindet und deshalb in der Tat eine einzige Herausforderung internationaler Forschungszusammenarbeit zum gegenseitigen Nutzen ist.

Selbstverständlich ist auch unsere Zusammenarbeit an essentielle Voraussetzungen gebunden. Natürlich nützt die beste Universitätsvereinbarung gar nichts, wenn nicht entwickelte inhaltliche Berührungspunkte vorliegen. Und, was nicht zu unterschätzen ist, man muß zusammenkommen und zusammenarbeiten, dann werden signifikante Beiträge möglich.

Ausgehend von diesen einleitenden Gedanken möchte ich zu drei Problemskreisen sprechen:

1. Zunächst möchte ich die Zusammenarbeit und ihre Entwicklung mit einigen persönlichen Eindrücken charakterisieren.

2. werde ich einige sehr allgemeine Fakten darstellen, die die gesellschaftliche Relevanz unseres Forschungsgebietes unterstreichen und damit den Beitrag der Zusammenarbeit unsere Universitäten zu internationalen Fortschritten auf diesem Feld kennzeichnen,
3. möchte ich einige kurze Bemerkungen zu laufenden Projekten und der perspektive unserer Zusammenarbeit anfügen.

Ich komme also zunächst zu einer persönlichen Reflexion der Geschichte des Brown - WPU - Austauschprogramms. Das ist auch deshalb notwendig, weil Prof. Kestin sehr bescheiden seinen eigenen Beitrag unzulässig verkleinert und gleichzeitig unsere Ergebnisse sehr hoch gewertet hat. Um die richtigen Relationen wiederherzustellen, will ich hier zunächst anmerken, daß man auf dem Gebiet von Präzisionsexperiment und exakter Theorie der Transporteigenschaften international von einer Eltern - , einer Kinder - und nun auch schon von einer Enkelgeneration spricht. Das gilt für Personen <u>und</u> Apparaturen. Ich meine, es ist ganz offensichtlich, wer hier die Eltern - und wer die Kindergeneration vertritt. Dasselbe weist ein Blick in die Literatur aus. So erfreulich also die Anmerkung unserer Rostocker Arbeiten ist, die sich sowohl in der Kooperation der letzten Jahre als auch im vorangehenden Teil dieses Beitrags ausdrückte, so sehr muß man die letztgenannten Relationen gerechterweise in Rechnung stellen...
Doch nun entgültig ein Blick in die Geschichte :
Als mein Interesse an thermophysikalischen Eigenschaften während meines Studiums vor etwa 20 Jahren geweckt wurde, und das gilt für alle Kollegen unseres Wissenschaftsbereiches Physikalische Chemie mit insignifikant kifferierenden Jahreszahlen gleichermaßen, waren mindestens 50% der zu

studierenden Literatur mit Name J. KESTIN, E.A. MASON, L. MONCHICK...verbunden.

Zu dieser Zeit begann unsere Gruppe unter der Leitung von Prof. F. OPEL ihr heutiges Profil herauszubilden. Dieser Prozeß war zwangsläufig beeinflüßt durch internationale Trends. Die Bestimmung von Virialkoeffizienten der thermischen Zustandsgleichung wird immer Bezug nehmen zu dem Standardwerk von E.A. Mason und T.H. Spurling, "The virial equation of state" /1/. die moderne Entwicklung bei Präzisionsviskosimetern auf der Basis schwingender Körper geht ebenso wesentlich auf Arbeiten von J. KESTIN und Mitarbeiten zurück, wie die Entwicklung der Hitzdrahtmeßtechnik zu einem primären Standard für Wärmeleitfähigkeitsmessungen /2/. E.A. MASON et al. entwickelten in den 60er und 70er Jahren sowohl die Theorie des Wärmeleitfähigkeitskoeffizienten als auch die der theorischen Transpiration /3.6/. Und nicht zuletzt gingen mit R.H. COLE /z.B.7/ auch wesentliche Impulse für die Untersuchung dielektrischer Virialkoefizienten von der Brown University aus. Unser Programm war natürlich nie eine einfache Kopie dieser Arbeiten. Es wies hinsichtlich spezifischer experimenteller Lösungen (Prof. J. Kestin hat z.B. auf die vorteilhafte, aber glastechnisch komplizierte Verwendung von Quarz als Zellmaterial verwiesen) und des an industriellen Bedürfnissen orientierten Substanzprogramms von Anfang an Rostocker Besonderheiten auf, die es gerade kooperationsfähig werden ließen. Hier ist nicht der Ort, ein solches Profil im Detail darzustellen, das wurde an anderer Stelle getan /8/. Ziel dieses Exkurses war es, nachzuweisen, daß Berührungspunkte zur Brown lange vor Unterzeichnung des Austauschprogramms bestanden.

Aber: wir kannten die Namen, wir wußten um die Adresse, und die war Brown University, Providence, u.s.w., Doch es war eben nur eine Adresse, auch das sichere Gefühl, daß dort eine wichtige Arbeit getan wird. Und noch einmal, aber: Es gab keinen persönlichen Kontakt, über Brown und Providence.

Selbstverständlich erwachte unser Interesse auf einem ganz anderen Niveau, als wir 1979 erstmals von der Universitätsvereinbarung zwischen unseren beiden Einrichtungen hörten. Sie bildete schließlich die Grundlage für einen Arbeitsaufenthalt von J. Kestin in Rostock im September 1981. Im Ergebnis dieser anregenden Diskussionen hinterließ er uns einen ganzen Problemskatalog und den Rat, mit der Beantwortung dieser Fragen experimentell Möglichkeiten und Grenzen der entwickelten Methoden exakt aufzuzeigen, weil das die unerlässliche Voraussetzung für eine effektive Mitwirkung bei der notwendigen internationalen Forschungskooperation sei. Natürlich folgten wir diesem Rat, sammelten Ergebnisse, leiteten Schlüsse ab. Nicht immer kamen wir gleich zu übereinstimmenden Bewertungen, aber auch hier war es der wissenschaftliche Meinungsstreit, der letztlich alle Beteiligten voranbrachte.

Unsere Resultate hatten uns Mut gemacht, und so beschlossen wir 1983/84 für die V. Rostocker Universitätstage im Februar 1986 eine Konferenz zum Thema "Makroskopische Eigenschaften von Gasen und zwischenmolekulare Wechselwirkungen" zu organisieren und international bekannte Kollegen zu Plenarvorträgen einzuladen. Wir waren sehr froh, daß alle unserer Einladung folgen konnten, darunter wiederum Prof. Kestin.

In der Rückschau bin ich der Ansicht, daß diese Konferenz den endgültigen Beginn der Kooperation unseres Wissenschaftsbereiches mit der Brown University und den Einstieg in die internationale Forschungszusammenarbeit markiert. Ich selbst konnte 3 Monate später erstmals die Brown University besuchen und gemeinsam mit E.A. MASON Probleme der thermischen Transpiration untersuchen. Die Resultate dieser Diskussionen bildeten einen Ausgangspunkt für Untersuchungen, die wir jüngst zusammenfassend publizieren konnten /9/. Aber mehr noch, dieser Besuch, verbunden mit der Teilnahme am "9th International Symposium on Thermophysical Properties", einem alle 3 Jahre stattfinden den bedeutsamen Kongreß auf unserem Arbeitsgebiet, war der Ausgangspunkt für unsere Integration in das wichtige Netzwerk internationaler Zusammenarbeit vor allem auf dem Gebiet der Transporteigenschaften, wie es J. Kestin bereits be-schrieben hat.

An dieser Stelle kommt auch die Gruppe von Prof. W.A. Wakeham vom Imperial College, London, ins Spiel, die gleichsam zu einem Koordinierungszentrum solcher Aktivitäten geworden ist. Lassen Sie mich schließlich anführen, daß auch mein Aufenthalt 1988 End- und gleich wieder Ausgangspunkt gemeinsamer Arbeiten war. So konnten wir auf dem "10th Symposium on Thermophysical Properties" in zwei komplementären Beitragen Resultate aus dem gemeinsamen Arbeitsprogramm vorlegen, das unter der Schirmherrschaft des schon genannten IUPAC Subcommittee on Transport Properties steht /10.11/. Auch das sehen wir, wie gesagt, im eigentlichen Sinne nicht als Endpunkt, sondern als Anregung auf höherem und breiterem Niveau zum gegenseitigen Nutzen weiter zusammenzuarbeiten. Diese Aussage beinhaltet natürlich die

Frage nach effektiven Formen wissenschaftlicher Zusammenarbeit: Nach meiner Ansicht können relative kurze Aufenthalte (und 1 bis 2 Monate sind gemessen an den zu lösenden Problemstellungen kurze Zeiträume), wie schon ausgeführt, nur Anfangs - und/oder Endpunkt längerer Zusammenarbeit zu vereinbarten Schwerpunkten sein, wobei das entsprechende publizierte und unpublizierte Material beider Gruppen an den jeweiligen Heimatuniversitäten weiterbearbeitet wird. Das setzt natürlich sowohl fortdauernde komplementäre Forschungsprogramme als auch regelmäßige persönliche Kontakte voraus. Dazu noch einige Gedanken am Schluß meiner Ausführungen.

Nun konnte man natürlich mit Recht fragen, welchen Gewinn bringt das Austauschprogramm dem einzelnen Wissenschaftler und seinem Kollektiv?

Das Seminar hat dazu viele wertvolle Gedanken zusammengetragen, ich denke an die Bedeutung der persönlichen Kontakte, des Redens miteinander und des Lernens voneinander, des Lernens auch über ander Länder und gesellschaftliche Systeme. Ich erspare mir weitere Wiederholungen, weil ich mit fast allen hier gefallenen Bemerkungen in dieser Richtung übereinstimme.

Ich möchte vielmehr einen wissenschaftlichen Gewinn anfügen, den sicher jeder für sich - natürlich in modifizierter Form - nachvollziehen kann, gleich ob er aus den Vereinigten Staaten zu uns nach Rostock gekommen ist, um hier zu arbeiten, oder umgekehrt: Die Zusammenarbeit mit den Kollegen in Providence, wie übrigens auch die mit jenen in London, hat mir vor allem neue Standpunkte insbesondere hinsichtlich der Relation von Grundlagenforschung und anwendungorientierter Forschung vermittelt. Im Zusammenhang mit dieser

Bemerkung muß man bedenken, daß ich an dieser Universität in einer Sektion studiert habe, in der Grundlagenforschung dominierte (was sich in den letzten Jahren gezielt geändert hat). In Providence wie in London kam ich zum Chemeingenieurwesen. Und der Gewinn einer solchen Kombination kann meines Erachtens gar nicht überschätzt werden.

Diese Bemerkungen bringen mich zu meinem zweiten Teil, in dem ich einige Fakten zur gesellschaftlichen Relevanz unseres Arbeitsgebietes und damit unsere Zusammenarbeit darlegen möchte.

Im Grunde genommen ist es ganz einfach: Industrielle Relevanz von physikalischer und/oder chemischer Forschung muß sich zunächst in gesellschaftlichem Fortschritt und nicht zuletzt in Form von Gewinn oder Profit (Geld) ausdrücken /z.b. 12, 13/. Das meint natürlich auch einen Beitrag zur Lösung anstehender globaler Probleme zu leisten. Es ist nicht schlechthin modern es: ist vielmehr dringend notwendig über solche globalen Probleme nachzudenken, zu diskutieren und vor allem zu einer friedlichen Entwicklung unserer Welt, zum vernünftigen Umgang mit der Biosphäre, mit Resourcen, zum sparsamen Einsatz von Material und Energie beizutragen. Ich denke, heute bestreitet niemand mehr, daß globale Probleme nur durch internationale Anstrengungen und internationale Zusammenarbeit zu lösen sind. Prof. Kestin hat das bereits an spezifischen Fragestellungen demonstriert. Ich will es aus der Sicht des Chemikers auf wenige Fragen reduziert und mit einigen Zahlen ergänzen: Jeder von Ihnen kennt mehr oder weniger das Aussehen einer Chemieanlage. Auch äußerlich wird sofort deutlich, hier werden Rohmaterialen durch riesige Rohrschlangen gepumpt, oft muß man sie erwärmen, manchmal kühlen, in der Regel mischen. Schließlich gelangen sie in den eigentlichen Reaktor, wo sie chemisch umgesetzt werden zu den gewünschten Endprodukten und mehr oder weniger zahlreichen unerwünschten Nebenprodukten. Dieses Gemisch wird abgepumpt, getrennt, das Endprodukt gereinigt u.s.w. Soweit die idealisierte Sicht auf einen Chemiebetrieb. Die Prozeßbedingungen im betrachteten Reaktor z.b. Temperatur, Druck, u.a.) hängen sowohl von den Anforderungen des chemischen Prozesses selbst, als auch, das wird manchmal unterschätzt, von ökonomischen Sachzwängen ab.

> Auf den ersten Blick scheinen sich an dieser Stelle die Wege von Naturwissenschaftlern und Ingenieuren zu trennen. Typische ingenieurtechnische Fragen wären z.B.:
>
> -Welchen Durchmesser müssen die Röhrensysteme haben?
>
> -Welche Wärmemengen muß man zu - oder abführen, um die gewünschten Temperaturen zu erhalten?
>
> -Welche Ausmaße müssen die Wärmeaustauscher haben, um diesen Energietransfer zu realisieren?
>
> ...und ähnliche Fragen mehr.

Die Antwort auf diese Fragen hängt natürlich in großem Maße von den stofflichen Besonderheiten der beteiligten Fluide ab, anders ausgedrückt von deren thermophysikalischen Eigenschaften. Für die Technik wäre es wünschenswert, <u>alle</u> benötigen Eigenschaften <u>alle</u> beteiligter Stoffe und ihrer Mischungen möglichst gleichzeitig und mit gleich hoher Qualität zur Verfügung zu haben. Für den Naturwissenschaftler, der sich der Grundlagenforschung verschrieben hat, lautet die Frage "Warum nehmen diese Eigenschaften für jenes Fluid gerade diese speziellen Werte an, und was können wir daraus an Verallgemeinerungen ableiten?" Der Physikochemiker würde diese Frage am liebsten zunächst für ein einziges möglichst einfaches System abarbeiten.

Die Realität ist, daß wir heute beiden Wünschen nicht gerecht werden können. Vielmehr zwingen uns ökonomische Restriktionen und die Mannigfaltigkeit der Natur, eine Brücke zwischen den Wünschen der Ingenieure und dem Wollen und Können der Wissenschaftler zu schlagen. Dir Formel lautet: Präzisions-experimente, verbunden mit der Weiterentwicklung der Theorie und entsprechender Modelle, müssen letztlich zu theoretisch begründeten Vorausberechnungsverfahren führen, mit denen wir Daten für Stoffe und Zustandsbereiche berechnen können, die uns - aus welchen Gründen auch immer - heute meßtechnisch (noch) nicht zugänglich sind. Das ist dann die Basis für effektive Computersoftware, die in Form von Datenbasen und Vorausberechnungmodulen entscheidend für ein effektives, material - und energieökonomisches Design von Chemieanlagen geworden ist. Aber um es nochmals zu betonen, der schnellste Computer mit dem größter Speicher setzt in jedem Falle die präzise Kenntnis von Schlüsselgrößen für Schlüsselsubstanzen in ausgedehnten thermodynamischen Zustandsbereichen, entsprechende Modelle und theoretisch gestützte Berechnungsverfahren voraus, andernfalls produziert er zwar Zahlen, aber keine akzeptablen Stoffwerte. In diesem Spannungsfeld ist unsere Arbeit angesiedelt, und das folgende, leicht nachzurechenende, inzwischen solch “klassische” Beispiel /13/ so demonstrieren, daß es eine praktisch unbegrenzte Herausforderung an Schöpfertum und wissenschaftlichen Eifer ist:

Unterstellt man nämlich einmal, daß für die chemische Industrie lediglich 15 reine Substanzen von Interesse sind und daß 10 thermophysikalisch Eigenschaften für alle möglichen

Mischungen bei 10 Temperaturen, 10 Drücken und jeweils nur 5 Konzentrationen benötigt werden, dann beläuft sich die Zahl der benötigten Messungen auf 10 Millionen! Nimmt man an, daß ein Wissenschaftler bestenfalls drei Messungen pro Tag schaffen kann, dann wird aus diesem realistischen Ansatz deutlich, daß es 10,000 sogenannte Mannjahre erfordern würde, um die Erfordernisse der Technik allein durch Messungen zu erfüllen. Stellt man in Rechnung, daß heute mehr als 6 Millionen chemischer Spezis bekannt sind und etwa 30,000 in der Industrie erzeugt und angewandt werden, wird die Problematik überaus deutlich. Schließlich ein letztes instruktive Beispiel, das für umweltfreundliche, energie - und resourcensparende Verfahren stehen soll:

"Methanolchemie" ist ein Begriff, der zunehmend für einen ganzen Zweig der chemischen Industrie steht, gleichsam auch als Inbegriff für eine "saubere" Chemie. Am Anfang der zahlreichen denkbaren Folgeprozesse steht natürlich der Methanolreaktor selbst. Im angenommen Fall werden Kohlenmonoxid und Wasserstoff zu Methanol umgewandelt und zwar in einem Prozeß, der das Vorwärmen des Reaktionsgemisches erfordert. Das wird realisiert, indem in einem Wärmeaustauscher der ankommende Strom der Ausgangsstoffe vom Produktstrom aufgezieht wird. Unsere Kollegen in London /z.B. 14/ haben in diesem Zusammenhang bei Modellrechnungen nachweisen können, daß die Verwendung zweier unterschiedlicher Tabellen für wesentliche Eigenschaften von Methanol mit international gleich hoher Reputationen bei der Auslegung des genannten Wärmeaustauschers zu einem Längenunterschied von 50% führt! Abhängig vom wahren Wert z.B. der Wärmeleitfähigkeit von Methanol bedeutet das im allgemeinen, daß der Austauscher überdimesioniert wird, wa zu erheblichen Meraufwand an Material, Energie und damit Finanzen führt. Ist andererseits dieser Teil des Reaktors unterdimensioniert, wird das Reaktionsgemisch nicht die notwendige Temperatur haben, damit sinkt die Ausbeute, und letztlich die Effektivität des Prozesses.

Das waren nur zwei Beispiele, die drastisch den Einfluß quantitativer Daten auf moderne industrielle Prozesse verdeutlichen. Ich bin sicher, daß auch der Nichtfachmann daraus ein Gefühl dafür bekommt, daß das nicht nur die schon erwähnte unbegrenzte Herausforderung an unser Talent, unsere Ideen und unseren Fleiß ist. Da keine Gruppe, auch kein Land, diese Probleme im Alleingang lösen kann, ergibt sich daraus zwangsläufig die Notwendigkeit abgestimmter internationaler Zusammenarbeit zum gegenseitigen Nutzen.

Unsere schon zitierten Arbeiten (Teil I) über die Korrelation und Vorausberechnung von Transportkoeffizienten sind Bestandteil dieses Programms. Ausgehend von Schlüsselgrößen und Schüsselsubstanzen trachten wir nach theoretisch begründeten Berechnungsverfahren, letztlich nach einer Weiterentwicklung der Theorie, die ihrerseits wieder zurückwirkt auf die angewendeten Disziplinen. Ein sehr schönes Beispiel für diese sinnvolle Einheit von Experiment, Theorie und Anwendung repräsentiert das Erweiterte Theorem der übereinstimmenden Zustände, auch bekannt als KESTIN-RO-WAKEHAM- (KRW)- Theorem /z.B. 10, 15-19/.

Literatur

1. E.A. MASON and SPURLING T.H.. "The Virial Equation of State", Pergamon Press, Oxford, 1969.
2. J. KESTIN, Phys. Chem. Earth 1981, 13-14. 195-319.
3. E.A.MASON and L.MONCHICK, J. Chem. Phys. 36 (1962) 1622.
4. L.MONCHICK, A.N.G. PEREIRA, and E.A. MASON, J. Chem. Phys. 68 (1978) 5277.
5. L.A.VIEHLAND, E.A. MASON and S.I.SANDLER, J. Chem Phys. 68 (1978) 5277.
6. E.A.MASON and A.P.MALINAUSKAS, "Gas Transport in Porous Media: The Dusty Gas Model", Amsterdam/Oxford/New York" Elsevier, 1983 and references therein.
7. R.H.ORCUTT and R.H.COLE, J. Chem. Phys. 46 (1967) 697.
8. G.OPEL and J.MILLAT, Wiss.Zeitschr. WPU Rostock, M-N Reihe, im Druck.
9. J.MILLAT, A.PLANTIKOW, D.MATHES and H.NIMZ, Z.phys.Chemie Leipzig, 269 (1988) 865.
10. F.J.URIBE, E.A.MASON and J.KESTIN, Physica, in press.
11. J. MILLAT and W.A.WAKEHAM, Int. J. Thermophys., in press.
12. J.KESTIN, "Thermophysical Properties in Science and Industry", in: Proc.Eighth Cymp. Thermophys. Props., (ed. J.V.Dengers), ASME, New York, 1981, Vol.I, pp.1
13. W.A.WAKEHAM, "A Transport of Delight", Inaugural Lecture delivered at Imperial College, London, U.K., 1st March 1988.
14. J.B.ARMSTRONG, S.F.LI., and W.A.WAKEHAM, ASME Winter Annual Meeting, 1982, paper 82-WA/HT-84.
15. J.KESTIN, S.T.RO and W.A.WAKEHAM, Physics 58 (1972) 165: J.Chem. Phys. 56 (1972) 5837
16. B.NAJAFI, E.A.MASON, and J.KESTIN, Physics 119A (1983)387.
17. J.KESTIN, K.KNIERIM, E.A. MASON, B.NAJAFI, S.T.RO and M.WALDMAN, J. Phys. Chem. Ref. Data 13 (1984) 229.
18. A. BOUSHERI, J.BZOWSKI, J.KESTIN, and E.A.MASON, J.Phys. Chem. Ref. Data 16 (1987) 445.
19. E.BICH, J.MILLAT, and E.VOGEL, Wiss.Zeitschr. WPU Rostock, M-N-Reihe, in press.

Anlage zum Beitrag von J.Kestin:
Ausgewählte Publikationen aus der internationalen Forschungskooperation des Wissenschaftsbereichs Physikalische Chemie der Wilhelm-Pieck-Universität Rostock

1) E.Vogel, E.Bich und J. Millate
"Untersuchungen zur Konsistenz von experimentalen Viskositäts - und Wärmeleitfähigkeitskoeffizienten und aus Edelgaspotentialen berechneten Werten."
z.Phys.Chem., Leipzig, 267 (1986) 131-144.

2) J. Millat, M. Mustafa, M. Ross, W.A. Wakeham, and M. Zalaf
"The Thermal Conductivity of Argon, Carbon Dioxide and Nitrous Oxide"
Physica 145A (1987) 461-497

3) J. Millat, M.Ross, W.A.Wakeham and M.Zalaf
"The Thermal Conductivity of Neon, Methane and Tetraflouro-methane"
Physica 148A (1988) 124

4) J. Millat, V.Vesovic and W.A.Wakeham
"On the Validity of the Simplified Expression for the Thermal Conductivity of Thijasse, et. al"
Physica 148A (1988) 153

5) E.Bich, J. Millat, and E.Vogel
"Interatomare Wechselwirkungen in Edelgasen und erweitertes Theorem der übereinstimmenden Zustände"
Wiss. Zeitschr. WPU Rostock, in press.

6) J.Millat, A.Plantikow, D. Mathes and H.Nimz
"Effective collision cross-sections for polyatomic gases from transport properties and their thermonuclear pressure differences"
z.Phys..Chem., Leipzig, 269 (1988) 865

7) E.Bich, J. Millat and E.Vogel
"On the He-He interatomic potential and related properties"
z.Phys.Chem., Leipzig, 269, (1988) 917

8) J. Millat, V. Vesovic and W.A.Wakeham
Final Report on:
"The Transport Properties of Carbon Dioxide Outside of the Critical Region"
IUPAC Thermophysical Properties Data Centre, Imperial College, London,
8pp.+60 Fig.s

9) J. Millat and W.A.Wakeham
"The Thermal Conductivity of Nitrogen and Carbon Monoxide in the Limit of Zero Density."
z.Phys. Chem. Leipzig, in press.

10) E. Vogel, T.Strehlow, J.Millat, and W.A.Wakeham
"On the temperature function of the viscosity of nitrogen in the limit of zero density."
z.Phys.Chem. Leipzig, in press.

11) J. Millat and W.A.Wakeham
"The correlation and prediction of thermal conductivity and other properties at zero density"
Int. J. Thermophys. in press.

12) J.Millat, M.Ross, W.A.Wakeham and M.Zalaf
"The thermal conductivity of ethylene and ethane"
Int. J. Thermophys., 9, (1988) 481.

13) J.Millat, V.Vesovic, and W.A.Wakeham
"Theoretically-based data assessment for the correlation of the thermal conductivity of dilute gases"
Ont. J. Thermophys., in press.

14) G.A.Olchowy, J.V.Sengers, V.Vesovic, W.A.Wakeham, J.T.R. Watson, and J.Millat
"The transport properties of carbon dioxide"
J.Phys. Chem. Ref. Data, submitted.

15) J.Millat
Chapter 10.3: "Thermal transpiration"
IUPAC Volume of Measurement of Transport Properties of Fluids, Eds: A. Nagashima, J.V. Sengers and W.A.Wakeham, submitted, December 1988

16) J. Millat, M.Ross, and W.A.Wakeham
"Thermal conductivity of nitrogen in the temperature range 177 to 270 K"
Physica, in press.

17) E.Vogel
"Konstruktion eines Quarzglass-Schwingscheibenviskosimeters und Messungen an Stickstoff und Argon"
Wiss.Zeitschr. WPU Rostock, Math. Nat. 21 169 (1972).

18) E. Vogel, E. Bastubbe und S. Rohde
"Kalibrierung eines Quarzglas-Schwingscheibenviskosimeters für Präzisionsmessungen an Gasen und organischen Dämpfen"
Wiss.Zeitschr. WPU Rostock, Math. Nat. 33 34 (1984)

19) E. Vogel
"Präzisionsmessungen des Viskositätskoeffizienten von Stickstoff und den
Edelgasen zwischen Raumtemperatur und
650 K"
Ber. Bunsenges. Phys. Chem. 88 89 (1984)

20) T. Strehlow, E. Vogel, E. Bich
"Bestimmung von logarithmischem Dämpfungsdekrement und Schwingungdauer aus Zeitmessungen für zwei unbekannte Winkelstellungen be einem Quarzglas - Schwingscheibviskosimeter"
Wiss.Zeitschr. WPU Rostock, Math. Nat. 35 5 (1986).

21) E. Vogel, L. Barkow
"Precision Measurements of the Viscosity Coefficient of Carbon Dioxide between Room Temperature and 650 K"
Z. Phys. Chemie, Leipzig, 267 1038 (1986).

22) E. Vogel, E. Bich, R. Nimz
"The initial density dependence of the viscosity of organic vapors: Benzene and Methanol."
Physica 139A 188 (1986).

23) E. Vogel, E. Holdt, T. Strehlow
"The initial density dependence of the viscosity of organic vapors: Cyclohexane and neopentane."
Physica 148A 46 (1986).

24) E. Vogel, T. Strehlow
"Temperature Dependence and Initial Density Dependence of the Viscosity of n-hexane Vapor"
Z.Phys. Chemie, Leipzig 269 897 (1988).

The Physics Cooperation between Wilhem Pieck University and Brown University 1978-1989

Hendrik J. Gerritsen

In the early years of the cooperation, Prof. R. Sändig visited Brown University to explore how the program might best be developed and discussed with the then Chairman, Prof. P. Stiles and the author, various aspects thereof. It was realized that various components could be identified and developed to mutual benefit, in particular:
1) Visits by scholars from Rostock to Brown for a duration long enough not only to get acquainted but also start some work of

interest to both physics departments. Brown's good library and computing facilities were additional assets.
2) Visits by scholars from Brown to Rostock during which some lectures would be presented and an impression gained about fields of possible cooperation. Such visits were sometimes combined with attendance at a conference in Western Europe.
3) Attendance of physics students from Brown at Rostock for one semester.
4) The start of some cooperative research after scholars became acquainted with each others' work.

These ideas came to fruition in the subsequent years and several scholars from Rostock have spent some time at Brown University such as Prof. R. Sändig, H. Ulbricht, W.D. Kraeft, H.J. Fitting, M. Schlanges and recently, shortly after the conference, Prof. G. Röpke. Also, advanced students U. Waechter and J. Sonnenburg recently spent a semester at Brown. Notable about these visits is that good contacts were made and that Prof. H. J. Fitting published (1984) an article with D. Parker, a researcher in Prof. P. Stiles group, Prof. Schlanges and Roepke identified and encouraged Brown Physics students interested in studying a semester at Rostock, and Dr. J. Sonnenburg has written an article on elasticity of polymers with Prof. J. Weiner with whom he worked several months during 1988.

Several scholars from Brown made visits accompanied by lectures, namely Profs. H.J.Gerritsen (optics, lasers), P.J. Bray (glass physics), M. Glicksman (plasmas in semiconductors). These visits contributed to the formation of plans for cooperation as well. Last year (1988) a start was made during the visit of Prof. H.J. Gerritsen to exchange ideas with Dr. Rueckholdt regarding the establishment of advanced optics labs for teaching purposes in particular in the field of interferometric optics in which both universities are interested.

Two physics students from Brown have taken a one semester physics course at Rostock, T.K. Vehse during 1981 and C. Barter during 1988-89. They were both very positive about their experience and praised the excellent teaching and care they received. Beforehand, some faculty contacts from both sides had helped both students with making good choices from the Rostock course offerings, so that both students were able to integrate the courses they took at Rostock quite well in their curriculum. In order to attract students into this program, it would help if an up-to-date booklet about teaching and research in Rostock's Physics Department would become available. This suggestion is presently under consideration for implementation by Rostock in the future.

It is of course rarer to find students at Brown fluent in spoken German that to find advanced Physics students at Rostock fluent in research English. Thus it is important that Brown students in Physics, who are interested in study at Rostock, take one or preferable two courses in German first. Motivation to undertake this demanding program would also be stimulated if such a brochure would become available and broaden the applicant pool.

As is hopefully evident from this brief sketch of the highlights of our ten year cooperation, detailed plans for cooperation have been developed and implemented and some joint research is beginning to emerge.

It is of course unrealistic to expect major successes in this field to be realized quickly or even in the long run. Even within the field of physics, major subdivisions exist and the strength and direction of these subfields is largely determined by historical factors, personal preferences of researchers and encouragement by respective national agencies or industries in particular of fields with technological connections. Nevertheless some useful exchange is taking place and it is expected that this will even further develop as time goes on. In particular in the next ten years the hope is to develop two areas that have not received attention. One is the exchange of professors to teach certain physics courses in the other country's university for one semester. The other area is to strengthen particularly complimentary aspects in research. A concrete example of this is to do research with methods developed in one institution on materials prepared in the other institution. Similar possibilities of complementing each others' research exist in theory where for example Brown's access to good computers might allow certain theoretical work done at Rostock to yield interesting results.

In finishing, I'd like to point out that aside from the scientific aspects of the projects, less tangible but nevertheless very important other results came out of the cooperation. Among these are the establishments of friendships and recognition of each other's environment. Such views are extremely important in a world in which powerful propaganda machines try to mold images which - when situations change - are then modified too.

The author hopes that the dangers of the arms race and the possibility that the cold war may lead to an unprecedented experiment in applied nuclear physics, will decrease. Getting to know and appreciate each other are certainly activities leading in the right direction which is to survival and a better world. In that respect too our joint program - very modest as it is- is making a positive contribution.

Student Voices

Dr. Bernd-Peter Smollich

First of all, please allow me to introduce myself. My name is Bernd-Peter Smollich, I am a doctor of medicine in the 3rd year of residency in Internal Medicine at the University Clinic. During the first year of my residency, I successfully completed my doctoral thesis under the supervision of Professor Klinkmann.

In 1985 I studied for one semester at Brown University as one of the very first medical students from WPU to go to the USA. Of course when I heard that I had been selected I was tremendously excited, pleased, but at the same time, the prospect was a little bit daunting. But after having spoken with many American students who had been to Rostock, I believe those feelings were perfectly understandable.

Overall, the warmth with which we were received was most touching and members of the Foreign Study Office (Prof.

D. Smith, Mrs. Belton), students, but most of all Prof. Empkie, gladly assisted us with all initial problems.

I was free to work out my individual academic program. The flexibility in the American medical curriculum was new for me, at first strange, but in time I learned to appreciate the possibilities that such a system afforded.

My time in Providence was taken up with a clerkship in community health, followed by two different clinical clerkships and finally I chose an experimental part in the laboratory of artificial organs under the leadership of Prof. Galletti.

The clerkship in community health involved study in a number of community health institutions. It provided a firsthand look at the American community health system, what the determinants of health care are, what resources exist and what the needs of the community are. That experience more than any other rotation taught me so much about medicine, not only its scientific aspects but its social aspects as well. It was a good beginning and was well suited to settling a final year medical student from abroad into his American surroundings.

The clinical electives focused on the diagnosis and management of patients. I had to provide initial patient work-ups and to follow in-hospital patients. I took part in formal teaching-rounds, pathology review conference, lectures and seminars.

It was a very busy but also absolutely interesting time and an excellent opportunity to come in contact with a wide range of patients and of course a good change to strengthen my language skills.

Before going to Brown I had just completed the experimental part of my doctoral thesis and at Brown I was able to make full use of the excellent library facilities and had access to a wide range of up-to-date literature.

As I have mentioned, I spent 5 weeks in the lab of artificial organs in addition to the clinical clerkships doing experimental animal research. This time was a good basis for my further research and especially for a six-month stay in Belgium where I worked experimentally and clinically and from where I just returned some weeks ago.

Talking about scientific research brings me to a point I want briefly to mention. I think to have a precise scientific topic in mind before going to Providence or going to Rostock, respectively, is useful. Besides the medical study in general it should be a real highlight to work on concrete scientific topics during this time, topics in which both universities are interested. The opportunity to go to Providence was definitely one of the most important events in my education. It was a great chance to see the country and to be introduced to the culture of the U.S. I formed a deep attachment to both people and places while in the USA and I believe that fascinating people are one of the most important resources the USA have to offer.

This exchange program is a wonderful possibility for students, for young people of both universities and both countries to gain a personal understanding of and feeling for a truly different medical system, for a truly different way of life.

Volker Ahnfeld

Before I really start, allow me to thank all those on both sides of the ocean who made my trip to Brown University possible and all the friends over there who made my stay as great an experience as it was. To name them all would take more than half an hour so this should do.

As I had said on the occasion of a Rostock-party: drinking vodka in Providence is not something that you dream of as a kid. But for me, being there was a dream coming true. The results of my stay can hardly be overestimated. This is not only true in terms of knowledge - it goes far beyond that. Of course, my knowledge improved enormously, not only as far as my scientific work is concerned - where the actual results will become obvious in the process of writing my thesis (I feel this work is getting off the ground) - but also what I know about the country, its people and culture, not to mention language command.

More importantly, I gained a much deeper understanding of and respect for other ways of thinking. In fact, these five months changed my personality. I am much more aware of how complicated life is, but also that people are still people. In this respect there is more than just a personal dimension. In this world there are more walls than bridges, especially in people's minds, but the number of bridges is growing, some walls will be torn down, and I will help it, as small as my contribution might be. that's why I wish to express my hope that the exchange program will go one, and (as we put it) at an ever-increasing level.

Joseph Polman

I remember when I was trying to decide in 1986 whether I would come as an exchange student to Rostock. I talked to several people who had returned from various programs in German-speaking countries and I was impressed by the enthusiasm of the ones from Rostock. I remember distinctly how one of them told me: "It will change your life...it will be the most important time in your life." At the time I couldn't really understand what she meant, and she couldn't really make it clear. But now, looking back on it, I realize it was true.

The fact that I'm back in Rostock for a second time is proof of how important it has been to me. After leaving here I felt a tremendous need to keep some contact with the people ad the place, and to continue working through all the thoughts and ideas my experiences here had spurred. Most of the other people who had returned from Rostock felt similarly, and at Brown a group of students who had returned from Rostock as well as students from Rostock and a few others who were just interested got together regularly. I also wrote my Comparative Literature thesis on individualism in modern US and GDR novels as a way of further working through an issue which interested me. But that wasn't enough - I made plans to come teach at the English Fachpraktickum last fall and when the opportunity arose for me to stay for a year of research, I jumped at it.

I'm interested in education and that interest stem from the hope that education can help to change and improve the lives of all involved. In other words, it is my hope to make the world smaller for others, as it has been made for me through my education. But looking at the state of education in the United

States, I notice so many problems. So many children fall through the system, so to speak: hours and weeks spent in school and they end up not even being able to read. This is an indication that the way we educate in the US is not necessarily the best way, so why not look at another, very different, system of education to see how they do it? That's what I'm doing by looking at education in the GDR during this year.

A great deal has been accomplished here: the children seem to finish school with a very good amount of general knowledge and very few seem to "fall through." They all know how to read, and I don't think I've been fooled into believing that. One of the first things I learned as I began to read after getting here was that, theoretically, societal change is necessary first before education can be as effective as it is here. In other words, to improve education in the US, I should go and start a socialist revolution! But that's not very easy, nor is that what I want to do. Rather, I want to try and learn as much as I can from the educational system here.

With that in mind I began trying to figure out the general system here: how it's set up and why, as well as its development. I found that it did not begin in 1945 with the Soviet occupation of the "Eastern Sector", as an American would be tempted to think, but rather had roots reaching back to the turn of the century and further. I also decided to look more specifically at foreign language teaching, which meant English since I don't know Russian.

I also noticed that some of my friends, who were studying here at the university to become teachers, had a very extensive teacher training - five years throughout which they take courses on educational theory, psychology, and teaching methods. The final year consists of practical teaching under the supervision of an experienced teacher. The length and depth of this training is a sign of the worth education is given here, as well as an acknowledgement of the importance and responsibility of a teacher in this society. In the US, it takes much less time to become a teacher.

When I visited school here, I was struck by the discipline and order in the classroom. Was I in a military school? Why did the pupils stand up, formally greet the teacher and report who was absent? And does this order stifle creativity? All these questions went through my mind, but another question proved key to a better understanding of these practices: where do they come from? The answer is that they're a part of the Prussian culture's conception of politeness and respect. And this isn't the only place in the world where schools and teachers have such formalities; I had just never seen them before.

Another subject which I came up against was ideology, and it is tempting as an American to see it as brainwashing in the classroom. There are plenty of examples of ideology's role in the texts and topics used in the English classes: unemployment, lack of housing, the worker's movement (most Americans don't even know there is one), and racism in capitalist Great Britain and the United States. Then I found out that many educators here are no longer satisfied with the "Schwarz-Weiss-Malerei". where socialist society is shown as completely good and capitalist society as completely evil, and value judgements are always made. They want to present a fairer and more realistic picture of the world. But ideology will remain in the classroom. Why? The answer cam in this revelation that all education is ideological, whether one says it is or not. And it seems to be that we in the US too often try to hide this. And what is worse, or more dangerous: to consciously present ideology and possibly debate and to do so, or to unconsciously do so?

I gradually began to learn about some of the problems here, too. Some teachers feel overworked; the curriculum in schools is somewhat rigid and does not give as much opportunity for individual choices as is prevalent in US schools; it is difficult for English teachers, for example, to find outside materials to supplement the textbooks. I came to the realization that nobody has perfection, and that we should all try to learn from and help each other, rather than constantly trying to prove that we are good and they are bad (no matter who "we" and "they" are -- they're interchangeable). This goes for the field of education, but even more importantly is a general truth.

I think that's the reason this Brown-Rostock exchange has changed so many lives; it spreads understanding. This occurs in the give and take of factual information, but even more importantly in personal understanding. And that is impossible for me to describe with words, because it is found in the feelings and friendships we have developed on both sides of the ocean, between real people trying to live in a world we all have to share.

Hopeful Texts and Despairing Readings? Re-membering Hölderlin in an age of revolutionary convergence

Duncan Smith

HÄLFTE DES LEBENS

Mit gelben Birnen hänget
Und voll mit wilden Rosen
Das Land in den See,
Ihr holden Schwäne,
Und trunken von Küssen
Tunkt ihr das Haupt
Ins heilignüchterne Wasser.

Weh mir, wo nehm' ich, wenn
Es Winter ist, die Blumen, und wo
Den Sonnenschein,
Und Schatten der Erde?
Die Mauern stehn
Sprachlos und kalt, im Winde
Klirren die Fahnen.

THE MIDDLE OF LIFE

With yellow pears the land
And full of wild roses
Hangs down into the lake,
You lovely swans,
And drunk with kisses
You dip your heads
Into the hallowed, the sober water.

But oh, where shall I find
When winter comes, the flowers, and where
The sunshine
And shade of the earth?
The walls loom
Speechless and cold, in the wind
Iron banners clatter.

The headlines of the International Herald Tribune of November 27 inform its readers that the "German Question" tops the agenda for the forthcoming summit off Malta. The Hamburg weekly, *Die Zeit*, for the same week and the weeks before proposes paths to German unity, eventually to German reunification. Chancellor Kohl offers a ten point plan to bring about the reunification of Germany in the rather near future and finds support for his views in the United States. Elsewhere, his opinions find little support - a vigorous rejection by the government of the GDR; skeptical condemnation by the Soviet government, icy calm rejection by the NATO partners, and silence by the Eastern European nations, busy with their own revolutionary changes. Later, the new GDR government makes its own proposals, proposals which Willy Brandt, speaking in Rostock, finds surprisingly and agreeably like the plans for a German Alliance which the SPD might also be able to support. Gorbachev seeks to make clear to a muddled European leadership that reunification is not a pressing agenda for him or his nation nor is it for the world of European politics. The stability of Europe, he argues, depends to a large degree on the success of the GDR in maintaining its sovereignty, and that success is the responsibility not only of the GDR leadership and the GDR people but of those in other nations who might be tempted to interfere.

Not in many many decades has the "German Question" topped any agenda in the West let alone been the subject of so much media coverage. The events which began in October with the exodus of so many GDR citizens through Hungary and Czechoslovakia and which culminated in the removal of the Berlin Wall, the opening of the borders to the West as well as to the East by the new Krenz/Modrow government in the GDR have brought the entire world to a re-examination of the "German Question". Experts on that question have emerged from every piece of woodwork.

Among the experts has circulated the question, "Why couldn't you have known, have predicted, the events in the GDR?" The answers have been varied in the nations of the West. In the far off USA, the question elicits hardly a pause in the theater-like media coverage of the "now events" of Berlin. History has left this place generations ago, banished by McCarthy and by his liberal colleagues, non-existent except as the history of great men and great men's deeds on most college and university campuses. The deep resonant bass of Henry Kissinger and the entire male chorus of political scientists and sociologists has intoned its belief that while this was known to be coming, the speed with which it came caught everyone off guard. Now, back on guard, the social scientists have recommended their data compilations and their predictions, fed into the presses and the wire services with monotonous regularity as the latest "truth".

In Western Europe, perhaps in Eastern Europe as well, the issue is less why "we" didn't know than what does this mean for our future. The deep resonant bass chorus of Anglo-French voices, supported by the mixed chorus of other and smaller nations west and east of the imaginary dividing line which has bisected Europe for forty years, intones instead its great concern that a reunified Germany could ... and therefore should not ...

Not absent from all of this hectic interest has been the conventual anti-socialist and anti-communist discourse. This ranges from the absurd in my local newspaper whose otherwise moderate newswriter of important events has concocted the usual nonsense about everything being "grayer in East Berlin" than in West Berlin, etc., to the comical in the large investments of pieces of the wall on sale in the USA as "souvenirs of freedom". The collapse of socialism - even its "failure" as announced by the SED after its extraordinary meeting in December, has been advertised in every written or visual piece of western journalism. The citizens of the West can presumably sleep soundly for the first time in well over one hundred and fifty one years now that the spectre which haunted Europe and subsequently the rest of the world has turned out to be only a bit of "undigested cheese" as Scrooge observed so sagely to the ghost of Marley.

It has been impossible to fathom the idea that a continuation of socialist revolution has occurred in the East while in the West the unsuspecting media lords may soon have to wrestle with the problem of having stimulated the sonamulent millions of poor and underhoused, undernourished Americans with images of people taking matters into their own hands. "If after all the East Germans can do it...!" We can well imagine that coverage will soon cease of these earth shaking events once the possibility of contamination begins to emerge from the cover of anti-communist rhetoric.

Lost in all this so far save in the FRG and the GDR has been the less strident discourse of the cultural community, especially that community of lesser lights who are not (yet) stars of the global media village. The FRG conservative weekly, *Der Spiegel*, of November 17 is full of the wisdom of Gunther Grass or Martin Walser and a picture at least of Christoph Hein. The literary opinion is sought out if the opinion comes from one of the proven greats. Truth does not so easily get located among those who are not already stars.

Re-reading Hölderlin's poem above, a poem with which I prefaced my brief talk at the Ahrenshoop Symposium in March, 1989, I find that both the poet and the reception of the poet as well as the poem remain "contemporary". I wonder why all of us concerned with GDR culture did not remark more prophetically on the increase of interest in so esoteric a poet and so strange a poetry in the GDR of the late 1980s. Without laying claim to accuracy of prediction or perhaps even being interested in the need for such predictions, it strikes me again that those who read and reflect on poetry are in a curious way perhaps more able to understand what must soon happen within a culture and within a state or nation.

When I gave the first version of this paper to a small audience at the March, 1989 Ahrenshoop Symposium, I sounded extremely guarded about the future of what has been a most successful exchange program. For the past ten years - (in fact, for the past seventeen years, for the program actually began in 1972 though it was contractually formalized in 1979) - there has been a small but steady change in the official political and often personal hostilities which are perhaps the underlying causes of all walls including that most famous wall in Berlin. Undergraduate students, graduate students, professors, and even administrators have crossed and in many cases re-crossed the official borders between mutually hostile or unfriendly nations and cultures in order to advance work in their own fields. In addition to the advancements undertaken and achieved in fields like early childhood diabetes prevention and cure, techniques for cardiac surgery, energy saving measures in theory and practice, new research areas in physics and chemistry alike, and the work done by Germanists, scholars of English, Education, and Sociology, etc., there have been the impressive experiences of personal understanding at all levels, especially perhaps among those undergraduates and few faculty members fortunate enough to have had longer stays in the other country, the other culture. Why then in the midst of such a record of achievement and success this appeal to caution in contemplating the "second half" of the life of this program?

All the right political signs have apparently been given. There is a generous attempt at re-structuring of thought and attitude on both sides of the walls each side has constructed. There is apparently more openness than ever.

Perhaps the impatience which increases in the presence of new possibilities accounts for the increased awareness of poetry like that of Hölderlin's. Despair may mount when victory seemed near only to fade like a mirage. And we may seek solace in beauty - the beauty of words perfectly formed to the expression of hope and despair.

I want to propose,however, in this essay a different explanation for the current interest in Hölderlin's poetry and in Hölderlin's life. I suggest that it is not only and perhaps not even particularly this poetry which currently accounts for our interest in Hölderlin. Instead, it is his silence - the other half of his life - which commands our attention.

What interests us is the utter relentlessness of the writer's insight into the necessity for his silence. Hölderlin became "silent" or a whisperer after the male hero of his life - Sinclair - was arrested - silenced on charges of high treason - speaking and writing against the dominant state and in favor of what amounted to a kind of revolution. Pierre Bertaux has tried in his book on Hölderlin to make a case for Hölderlin's relative sanity, ascribing the acceptance of the "myth" of insanity to the drama which such myths make for later reception - scholarly as well as more popular reception - of a writer's works. My colleague and friend Dmitri Michailovich Urnov, a noted Soviet literary scholar, once answered my question about the relative diminution of the Soviet poet Yevgeny Yevtushenko by saying that the Russian readers like their poets to die young - preferably of suicide or alcohol. That makes them and their work acceptable. Hence Vysotsky became a second Yesenin to an enormous number of Russian readers and worshippers of "drama."

There is clearly an increased drama of such silence when it is imposed by suicide or insanity. It renders the writer and her or his works somehow more credible. They have penetrated more clearly into the abyss which is human society, human life. They have seen clearly - too clearly, and have realized with Camus that suicide is the only existential question.

Whether Bertaux is correct may remain a large question. Social pathologies, sometimes supported by otherwise less evident biological pathologies, certainly have an effect on their "sufferers" similar to the effect of a "pure" biological pathology. Where such pathologies and their symptoms are more an effect than a cause - such as German society and culture at the turn of the 18th to 19th Centuries - one thinks instantly of Werther and the wave of suicides allegedly sparked by reading that novel - the sufferer may not be legally or pathologically "mad." But *vis a vis* the dominant "mad" culture, she or he will have to accept that judgement and abide by the treatments accorded such pathological symptoms. Suicide is a means to escape the punishment of the living body. Burial in unmarked or unconsecrated graves, dissection even, might have been the fate of the body, but the "soul" or spirit had long since fled.

For those who went insane, there was treatment of a kind. If the insanity was clearly of a biological source - as with Nietzsche or Lenau who suffered from the final stages of a venereal disease - there can be no doubt that the insanity is not in the first place socially derivative (though obviously arguments are made about the nature of sexuality and of medical knowledge at a given time in a society which might have co-produced the disease.)

Bertaux argues that Hölderlin adopted a successful strategy to allow him to remain alive but freed from the twin burdens of social life -"normal" social life, that is, and from the burdens of poetic composition which represented or even was a part of the increasingly "disturbed" realty of early 19th Century Europe.

Bertaux's thesis is most convincing to me when I read it in this way. Hölderlin's "insanity" becomes a strategy of resistance which takes the form of a social pathology with mild to moderate dysfunctional symptoms. Hölderlin cannot live on his own. He cannot do without immediate support both financial (from his mother and after her death from a trust she established with his father's money) and physical - the care he receives from the owner of the house in which he resided during the entire second half of his life. He can and does receive friends, take walks and excursions into the countryside. He can and does continue to write, though what he writes does not command much poetic interest except for those who are students of his entire work and his life.

Insanity then was Hölderlin's state. He may have been, after all, what we today refer to as a "manic depressive or a bi-polar depressive", terms as vague in their description of what state a sufferer from this affliction may be in as they are frightening. It was moreover his "strategy more than his "state." By means of this strategy Hölderlin managed to avoid dealing with the outside world in any 'responsible" way. Given his grief over the death of Susette Gontard, his shame and guilt that he could not give her the love she offered him; given his terror at being a male when those males - especially one of them, Sinclair - whom he admired were in desperate trouble, Hölderlin's strategy was a successful one. At a relatively early age in a relatively long life, Hölderlin managed to abandon both the appearance of dealing with grief and shame and guilt. He also managed to abandon the necessity of "being a man," otherwise the often recommended strategy for those born male who cannot, however, conform to the dominant male behavior patterns from want, supposedly, of a powerful enough early role model, the lacking or missing father.

The same strategy put an end to those incredibly stressful wanderings through Germany and France by which Hölderlin may have been trying to flee from the consciousness of those internal problems. The descriptions of Hölderlin's physical state at the end of his wanderings suggest to me nothing so much as those homeless men of our own urban streets who walk so purposefully and so incessantly through their days, the autistic action of walking their most powerful sedative.

And there was the political situation of his "Fatherland" as well as of all of Europe, a situation to which he was acutely attuned. Class and gender, already notions of oppression for him, could only be transformed into poetic imagery for a given period of time. Poetry as action was already a conception. Yet it was a time when the results of the other forms of action were producing in so short a period of time the tangible results of apparent political and social liberation closely followed by the even more tangible results of political reaction. These included arrests, deportations, imprisonments, torture, and executions - the entire gambit of male interaction with the world males produce and command. Poetic action, not undangerous itself as Hölderlin suspected or knew, seemed a pale form indeed in such a time, a means of last resort, a further recognition of the great despair which pervaded those times, at least among the marginal figures of the culture, writers, poets, artists, and other traitors to a dominant culture.

Hölderlin had come to understand that political situation initially as a kind of romanticism in which images of ancient times mingled with contemporary battles and wars of liberation such as that to which Lord Byron went in Greece. Yet despite his sympathy with the cause of change, of reform if not revolution, Hölderlin did not take up arms on behalf of any part of the struggle, nor does he seem to have participated much in the councils of reform or revolution. His apparent terror at learning of Sinclair's arrest on charges of high treason for which a long imprisonment or even death was a normal state response, are understandable. The sensitive poet was not about to take upon himself the physical burdens of exposed marginality. His internal and external lives were hard enough.

This is not, I think, physical cowardice, though that is not itself an unintelligent strategy for dealing with a pathological social formation such as the male state of 19th Century Europe. Instead, it reflected awareness of the limited scope for resistance. The State or the dominant social formation which included the religious organization Hölderlin was always about to serve and always able just barely to escape, was "no mean opponent." The state also included the family, his own with the dominant mother in the father-less household, as well as the families he served as, of all things, a teacher, - together with the pastor the ultimate figures of cultural authority within the bourgeois family.

This state as well as any perceived resistance to it was entirely male, and Hölderlin was ill-equipped by circumstances of his own family to deal with that fact. Resistance was necessary but the paths to resistance involved acceptance of the forms of domination and poetry as the only proven means of escape. The twin roles of tutor and poet - so often found together in exactly this time of history and often in our own time - seem paradoxically contradictory in their social meaning.

Hölderlin had good reason to adopt some strategy of resistance when eventually he was no longer able to continue the conflict between reality and the ideal world he himself so often depicted as a representation of what that world could be. Insanity - real or fabricated - was his chosen mode of resistance. Imagine the paradox. Hölderlin, the poet of extreme and tender sensibility, is insane. The world of the European patriarchal state, replete with violence done to everyone and everything, is sane.

Small wonder that this life as well as this poetry has found resonance today.

Constant hostility to the dominant order produces cynicism in those who feel the obligation to resist. It can also produce self-hatred. At some level we recognize our complicity. It may be as determined as the complicity of gender. Those who are sensitive males and who resent and loathe the treatment of women or the production and reproduction of gender, remain males. They are doomed to a biological complicity. It may be class complicity. It may be racial complicity. To avoid the self-hatred which may result from too many clear insights into the nature of oppression, we may celebrate the aesthetic. We may also seek identification with non-violent strategies of resistance, however terrible and tragic in their consequences. Hölderlin's life and his writings offer both.

I wrote above that this present "era" of political accommodations - good will between East and West - brings with it the apparent advantages of all such periods. These are peace, freedom from the worry of global conflagration, plenty or at least more, and the opportunities presumably to develop our humanity by using resources previously expended on destroying that quality.

Yet curiously the interest in strategies of resistance has risen. Many suspect that all is not well and that the very accommodation may be symptomatic of a greater crisis. That greater crisis may be the development of a society and culture without alternatives - without alternative political or social discourses, without alternative places and spaces in which to find relief from the totality of convergent western material culture. After November 9, 1989, the situation may paradoxically if necessarily have reached truly crisis proportions.

It is clear that environmental issues have risen to the forefront of much of our consciousness and that we - not "they" - have begun to do something about those issues. The consequences of our previous waste have become noticeable at many levels of ordinary existence. European and Amazonian forests are ravaged, perhaps beyond repair, the world's waters are polluted even at their farthest reaches. Oil pours over the water and the land, and the destruction of non-human species in the interests of developing our resources for human existence progresses ever faster.

Such issues may tend to unite us. I noted in my last visit to the Soviet Union a year ago, that there was much talk of the wasteland which has been made of the vast supposedly untouched reaches of that country. True or not, we have become conscious of these issues in all our nations, though "they" - not we - have the actual power to continue to despoil or not. We - those of us living in the midst of plenty - have the right to be complicit - or not.

We are also becoming aware of the absence of differences, of the growing homogenization of our world. The apparent abandonment of socialism has lessened not increased the possibilities of resistance. It may well be the case that the abandonment of such differences as did exist between socialism and capitalism is entirely in the interests of socialism from a tactical point of view, but at the moment it looks more like the removal of one of the most important statements of potential no matter how troubled the actual state of that potential may have been.

The reasoning behind this sentiment - and it is a purely speculative sentiment - comes from thinking about the ways in which the world has become a "global village" over which certain forms of control are exercised - some necessary to the survival of the village and some necessary to the survival of those in control.

I do not trust the instinct of total homogenization which brings with it the removal of those differences on which too hope for alteration rests. Ecology will benefit me - us - little if it only saves the whales and the oceans and forests but does not save the possibility of emergent differences, the "movement" of the dialectic itself and less abstractly the refuge of hope when the new totalitarian consumerist culture asserts its complete sway over peoples' borderless lives.

Even if this speculative sentiment about Hölderlin's reception today were to be accurate, is the model thus offered us - retreat into a tower, silence, melancholy, survival - a viable one? Can or should it attract us or does this suggest a pathology among those so attracted? If Bertaux were correct in his assertion that Hölderlin was not insane, is such retreat a useful form of resistance?

We are accustomed to more heroic models. The "lords of the global village" have further proliferated this manner of resistance by giving us images in print and pictures which tell us that resistance is only that when it looks a certain way. That certain way is still often enough individuals or movements engaged in desperate and obvious struggle with the world which they overcome by force of will and a wide variety of weapons of violence.

On the level of psychological resistance we are perhaps more fortunate. We do have some images of resistance taking other than such heroic forms. I am thinking here of all the writings by women or images, usually by women in the GDR which trace the discovery of their oppression in even the allegedly most liberated of social states. The discoveries are still the dominant themes, rather than being replaced by examples of "heroic resistance." Iin the re-memberings of texts by women from past times - the writings of Günderrode, for example - stress is laid on the charting of the discovery of oppression, of the "reality" of the social constructs in which she and other women lived. Such heroism as may actually have existed lies more in the documentation, the willingness to find out than in the ability to overcome.

A postcard I purchased years ago in Berlin (West) depicts a blank house wall on which are painted the words: *Widerstand braucht viel Phantasie,* (resistance requires a lot of imagination) - a cliche which is not yet really that because its validity has hardly been overcome by its widespread acceptance. In an era of the "global village" and global integration where alternative models for social and community living are steadily eroded by the processes of accommodation and convergence, it is difficult to imagine what resistance does look like and why, indeed, it is necessary. The removal of the "wall" and the intention to obliterate differences in ideology between the West and the East is also the imperative call to the powers of imagination to construct resistance - soon, fast!

Reading Hölderlin's earlier poetry, the work of his "sane" periods, I am struck by the emphasis given the heroic. He lived in heroic times, after all, when the world was undergoing a fundamental political and social reorganization which promised to expand human consciousness. The images from the received world of antiquity, the emphasis on the gods, the adoption of the metrical systems of the classical past - all these things speak to me of a stretching of the poetic and personal self beyond any boundaries of reality which Hölderlin could have known. The recurring images of the *Heimkehr* - that idealized state of returning to the homeland, meaning the very local and regional homeland including an idealized family which in its reality appears to have tyrannized him, are dominant in much of his work at this time. The combination of (poetic) life among the gods and (poetic) life among the idealized family required not only the energy of poetic conception. It must have required as well an enormous personal energy which exceeded the boundaries of what Hölderlin could actually do.

Added to this stress which began for him at an early age with the composition of poems extolling tyrants, was the stress of his relationships to men and women, his "love" relationships. Deprived of fathers twice, he experienced in the harsh atmosphere of the schools he attended all the sadism and masochism of the lonely male being turned into a "man." He sought and seldom found any solace in female company, idealizing the one major relationship - that with Susette Gontard, a conveniently married woman - while fleeing from its very threatening physical consequences. He never achieved separation from his mother, that suffering figure of domination on whom he appears to have been utterly dependent all his life for his physical sustenance and for his psychological nurturing and who failed him conceivably in both roles.

The homoeroticism of his relationships to men - the search for the father figure is an appropriate designation for some of those relationships - was equally muddled in the very confusion of wants he experienced in this sector of his life. He had no professional success, no reflection back onto himself of any externalized worth. He failed as a private tutor, a pastor-to-be, a poet, and even a librarian. As far as I can see, nothing worked well in any sphere of his life until he became aware of his own personal reality and "retired" to his tower and to "insanity," conceivably the sanest decision he made in his entire life. It could have mattered little to the older Hölderlin that subsequent generations would admire his poetry and his life, and respect him and his work as a model for their own resistance.

I am now bypassing any other stated reasons for the present reception of this poet and his works in Europe and especially in the two German states and concentrating on this single speculation. Hölderlin demonstrates for me an example of the phantasy required to exist and resist.

His poetry is full of this resistance even when it appears to be an accommodation with the conventional admiration of the gods and goddesses, of the fatherland or homeland, or of "love." In these cases the resistance may lie in the dedication to the use of the most complex poetic forms. They may also have been the then prevalent poetic conventions as well, but their employment by Hölderlin eventually exceeds that limitation. They are devices of estrangement which often allow - and this I find the most unique thing about his poetry - a single line or short passage to stand out in a beauty of expression which is "not normal" for those times or our own times. He is an extraordinarily modern poet as has been widely noted. His awareness of estrangement and of the power of language to conceive and dispel estrangement in a dialectic of expression is more often encountered in late 20th Century writers. Even his last poems - those from the period of his "insanity" - often contain such formulations which may merely be our way of reading, of re-membering Hölderlin.

Resistance is encountered in his work from the perspective of the very late 20th Century industrialized culture in which I read and re-member.

The apparent torment of his life, the romantic reception accorded the "mad" poets of all times, is less appealing as a strategy of resistance. Indeed, the media lords of the global village have made this image their dominant one, accepting and promulgating the received image of poetic genius which finds its apotheosis in the images of Mozart - child genius dead at thirty five and never understood or appreciated. Milos Forman's re-membering of Mozart and Salieri is but the most blatant example of this preferred reception. Mozart's early death has contributed too much to the legend of his genius. His music - and had there been more it would have been wonderful too - needs no such glossing to be heard, though it is now impossible to hear it without those glosses.

Hölderlin's tower represents a moment of accommodation with actual reality. He was not a "prisoner" in the tower. Nor did he live in a tower at all. He lived in the house of a plain citizen of Tübingen who cared for him in some of those ways which he needed and at the expense of his deceased father who had left money to the family which could obviously be used in crises. He lived after 1806 as he could live rather than as he once supposed himself capable of living. Such accommodations with reality are disappointing to many who prefer the glamor of insanity and the continuation of the former self.

What else, after all, could he have done? He could not resolve the difficulties with his mother, with his homoeroticism, with the women he wanted to love and succeeded in only hurting. He could not persuade the world that he was a "great poet." He could not physically fight the visible or invisible tyrants. He could write poetry, but even that was denied him when the poetry became enmeshed in the themes of his life which remained outside of his ability to alter or understand.

Probably we all prefer the heroism of Kleist who took his own life after blowing out the brains of a co-dependent woman he allegedly loved. It is not for me to judge which of the two paths are "right," for neither of them seem happy paths, though Hölderlin's is at least non-violent.

We may have reached a stage in our history when resistance to what we perceive to be a tyranny can only take the form of retreat. Hölderlin is a model for such retreating, a painful one because it is linked to our received and constantly promulgated images of heroic resistance against which it appears at least "unmanly" and certainly "insane." The sane man would do something. And that is precisely the problem. He would and the results are calculable in terms of the human cost which results or at least has thus far always resulted when "sane" men "do things."

Now I want to return to his poetry - the poetry which has excited so much praise in his own times though not from the great "geniuses" Goethe and Schiller, and which today is once again a center of European, indeed western poetic creation. It is impossible to separate that strange and apparently so sad life from these poems and from the Hyperion novel. The life moves us at once. The poetry moves us today only as we become immersed in the extraordinary lyricism of Hölderlin's language art and in the rhythms and meters which are the poet's field of aesthetic and political action. The beauty of the verses is further heightened by the abrupt silence which evokes a response similar to that I feel, for example, when Schubert's unfinished symphony is played or when a Furtwängler or Solti conduct a Beethoven symphony mediating its exceptional sound with moments of silence so long as to provoke in me an almost unbearable longing - for more, for continuation. When Hölderlin stopped writing, the effect was and is the same.

Hölderlin's poetry and the reception of it - of late especially in the German Democratic Republic - is the poetry and the reception of an admonition to alter the world by altering the self which constructs that world. It is a political poetry of the highest kind, formulated in those meters and rhythms which concentrate the energy of their content onto the charge: "liberate yourself!" This charge, literally delivered on several occasions in Hölderlin's poetry, to the Germans is now being transformed by "East" Germans into actions viewed even in the ideologically hostile West as truly revolutionary beyond perhaps the superficial description of them as "the overthrow of their communist masters."

<u>An Die Deutschen</u>

Spottet ja nicht des Kinds, wenn es mit Peitsch' und Sporn
Auf dem Rosse von Holz muthig und gross sich dünkt,
Denn, ihr Deutschen, auch ihr seyd
Thatenarm und gedankenvoll.

Oder kömmt, wie der Stral aus dem Gewölke kömmt,
Aus Gedanken die That? Leben die Bücher bald?
O ihr Lieben, so nimmt mich,
Dass ich büsse die Lästerung.

To the Germans

Do not laugh at the child when with his whip and spurs
Brave and mighty he feels up on his rocking-horse,
For, you Germans, you too are
Poor in deeds though you've thoughts enough!

Or as lightning from clouds, out of mere thoughts will deeds,
Potent, come leaping out? Books now begin to live?
O, my dear ones, then seize me,
Make me pay for my slanderous words.

(translated by Michael Hamburger)

Had our symposium been held less than a year later, this might well have been the poem - this and its later, longer version which I will cite below in this article - which prefaced the presentation.

The reception of Hölderlin in the GDR, the response of readers to the poet's works and to the poet's life, has included the actions which have led to a renewal of the revolution in the GDR. Some commentary has made mention of the fact that these events in the GDR are the first evidence in modern German history of a revolution from below, perhaps the most astonishing fact in German and European history since 1848. It has not been only a rebellion against the leadership of the SED though it has been that too. Like all revolutionary movements, the events in the GDR possess their own momentum. They are creating their own futures and overturning them, seeking to establish those forms of governance with which a socialist people can live. "Imagine!" said a woman speaker at a Berlin rally lately "A socialism you can love!"

We confront a future balanced between the poet's words: a second half in which the poet and we hear the *Klirren der Fahnen* or a second half in which the poet gladly pays for his slanderous words.

In an article in the Times Literary Supplement on lyric poetry in Romania, an English critic remarked that poetry seems to flourish wherever oppression is at a peak. There seems to be an inverse relationship between the desire to speak in lyric formulation and the inability to speak otherwise. Such a circumstance must gradually make only speaking in western society an impossibility and should lead to each of us being a poet, for the oppression of the media world by the lords of the global village has reached ominous proportions. In the GDR we might confidently expect a diminution - should we accept this rather simplistic formula - of lyric poetry, a glorious laying to rest the re-membered poet and the re-membered poetry. Not then in sadness would the poet leave this stage as he indicated in that longer version of the poem An Die Deutschen:

An Die Deutschen

Spottet nimmer des Kinds, wenn noch das alberne
Auf dem Rosse von Holz herrlich und viel sich dünkt,
O ihr Guten! auch wir sind
Thatenarm und gedankenvoll!

Aber kommt, wie der Stral aus dem Gewölke kommt,
Aus Gedanken vieleicht, geistig und reif die That?
Folgt die Frucht, wie des Haines
Dunklem Blatte, der stillen Schrift?

Und das Schweigen im Volk, ist es die Feier schon
Vor dem Feste? die Furcht, welche den Gott ansagt?
O dann nimmt mich, ihr Lieben!
Dass ich büsse die Lästerung,

Schon zu lange, zu lange irr ich, dem Laien gleich,
In des bildenden Geists werdender Werkstatt hier,
Nur was blüht, erkenn ich,
Was er sinnet, erkenn ich nicht.

Und zu ahnen ist süss, aber ein Leiden auch,
Und schon Jahre genug leb' ich in sterblicher
Unverständiger Liebe
Zweifelnd, immer bewegt vor ihm,

Der das stetige Werk immer aus liebender
Seele näher mir bringt, lächelnd dem Sterblichen

Wo ich zage, des Lebens
Reine Tiefe zu Reife bringt.

Schöpferischer, o wann, Genius unsers Volks,
Wann erscheinest du ganz, Seele des Vaterlands,
Dass ich tiefer mich beuge,
Dass die leiseste Saite selbst

Mir verstumme vor dir, dass ich beschämt
Eine Blume der Nacht, himmlischer Tag, vor dir
Enden möge mit Freuden,
Wenn sie alle, mit denen ich
Vormals trauerte, wenn unsere Städte nun
Hell und offen und wach, reineren Feuers voll
Und die Berge des deutschen
Landes Berge der Musen sind,

Wenn die herrlichen einst, Pindos und Helikon,
Und Parnassos, und rings unter des Vaterlands
Goldnem Himmel die freie,
Klare, geistige Freude glänzt.

Wohl ist enge begränzt unsere Lebenszeit,
Unserer Jahre Zahl sehen und zUahlen wir,
Doch die Jahre der Völker,
Sah ein sterbliches Auge sie?

Wenn die Seele dir auch über die eigne Zeit
Sich die sehnende schwingt, trauerend verweilest du
Dann am kalten Gestade
Bei den Deinen und kennst die nie,

Und die Künftigen auch, sie, die Verheissenen
Wo, wo siehest du sie, dass du an Freundeshand
Einmal wieder umarmest,
Einer seele vernehmlich seist?

Klanglos, ists in der Halle längst,
Armer Seher! bei dir, sehnend verlischt dein Aug
Und du schlummerst hinunter
Ohne Namen und unbeweint.

To the Germans

Never laugh at the child, seeing the silly one
Feel important and great up on his rocking-horse;
O my brothers, we too are
Poor in deeds though we've thoughts enough!

But as lightening from clouds, out of mere thoughts perhaps
Will the deed in the end, lucid, mature, leap out?
As from dark orchard leaves, from
Quiet scripts does the fruit ensue?

And this hush in the crowd, is it the joy before
Joy's occasion? The awe marking the god's approach?
O then seize me, my dear ones,
Make me pay for my slanderous words.

Far too long now, too long like a poor layman I
In the shaping spirit's workshop have roamed perplexed;
Things half-grown I can see there,
What he schemes I can not make out.

Sweet it is to divine, but an affliction too,
And enough years I've spent loving as mortals do,
Doubting, uncomprehending,
Ever moved in his presence who

Ever closer to me out of his loving soul
Brings the constant great work, smiles at the mortal man
Where I falter, and to its
Ripeness brings the pure depth of life.

O creative one, when, genius innate in us,
Wholly will you appear, soul of our fatherland?
So that lower I bow then,
Of my strings the most muted then

Dare not sound, and ashamed, dumb before you I droop
Like a flower of the night, heavenly day, and long
But to wither with gladness,
When all those in whose midst I could

Only mourn- when our towns, brightened now, are awake,
Open and communal, full of a purer fire,
And the mountains of German
Lands are mountains the Muses haunt

Like those glorious ones then, Pindos and Helicon
And Parnassus, and here under the fatherland's
Golden sky the pellucid,
Free and enlightened gladness gleams.

True, the span of our lives briefly extends; we can
See and count the few years granted to us on earth,
But the years of the peoples,
These what mortal man's eye has seen?

Though your soul roams away, winged with its yearning soars
Far beyond your own time, mournful you linger here,
Cold on desolate shores, with
Your own kind, but estranged from them;

And those others to come, those for whose advent we wait,
Where, O where can you see them, that once more you'll be
Warmed by one hand that's friendly,
Audible to one living soul?

Without resonance, long empty for you it's been
In your hall, poor seer, now; yearning your eye grows dim,
And you drowse away, vanish
Never noticed, unnamed, unwept.

(translated by Michael Hamburger)

Afterword

Duncan Smith

There are so many times in our lives when we need a kind of steadfast optimism. We need it as parents and later, when our own parents become elderly, we need it as children. Perhaps it would be equally true that as children we need it just because we have parents

So too in our larger lives, our political lives we need that optimism often. I would propose that we are again - some would say still - living in just such a time.

For almost two decades there has been an exchange of people through what used to be considered by "both sides" an almost impermeable barrier to just such exchange. Even before the establishment of official "diplomatic" relations between the United States of America and the German Democratic Republic, faculty members and graduate students from Brown University and the then Rostock University (later to become the Wilhelm Pieck University) were beginning to know each other.

It is difficult, especially now in the immediate aftermath of the events leading to the revolutionary changes in Eastern Europe and the Soviet Union to remember that there ever was a time when contact between our two cultures, our two societies was tremendously difficult, even precarious.

Yet in 1972, when the first contacts were made between faculty members of the German Departments of the two institutions, contacts facilitated initially by a classicist - (there will be reason to remark on the significance of this origination later) - a faculty member from Rostock visiting Columbia University in the very dark days of 1969 - the apparent rareness of such contacts was very plain to almost everybody. To be a citizen of the GDR - “East Germany”, the “Zone”, the “other Germany”, or the “so-called GDR” as the West German press in particular liked to call the GDR - and to be in the West, in the USA in particular - was to invite comments like “Are you going to stay?” “You aren't going back, of course?” It seemed impossible to imagine in the West that a citizen of the GDR would be in the West at all unless she or he were a “refugee”.

There were many understandable reasons for this western reaction. There were many GDR refugees in West Germany, citizens of the GDR who had left their country because they could not accept that country's ideology or the material conditions of their lives lived within that ideology. So many people left the GDR by 1961 that the country decided on the extraordinary step of erecting an impassable barrier - the wall in Berlin - to make such emigration not only illegal but virtually impossible.

The reasons for the wall have been presented by each side and rejected by each side - that is, each side has rejected the other's explanation. “Schutzmauer” or “Schandmauer” - as so often, which side you were on depended so much on which side you were on.

Exceptions to the dominant ideological representation of the wall by citizens of the West were rare. It was far more common understandably for citizens of the East to dissent from their government's position. This dissent seemed to prove the West's superior understanding of the "true" motives for the erection of the wall.

This in turn made contact with the GDR even more difficult. It was understood in the West that such contact was unnecessary, even immoral, since it could only shore up an illegal and immoral regime which operated against the will of its own citizens. It was unthinkable that an alternative point of view could exist, one which, western enough, sought to escape from the ideological dualism of "good and bad", "wrong and right", "West and East".

In short, it was unthinkable that a Westerner could have any positive impression of the GDR unless he or she were the "victim" of brainwashing, that wonderful McCarthyite term which permeated the American consciousness in the aftermath of the Wisconsin Senator's reign of terror.

The view of the West in the East was equally understandable. There the representation of the GDR by the West, by the West Germans in particular was so comprehensively negative that those who did not actively seek to leave their country, regardless of their own personal attitudes toward all the policies of their leadership, found themselves with one thing, at least, in common with that government. They resented and found threatening the posture of those western governments which questioned the legality and morality of the

GDR. This was even more the case given what was known about the West and its own legality and morality.

The representations of the West, partly ideologically motivated and partially defensive propaganda, gave rise to a dim view of capitalism, of western mores, and of the chief western nation in particular. The United States of America, whatever else it might be, was clearly no friend of the poor in its own country, of the poor and downtrodden in other nations, and when provoked, would attack and destroy those who sought to alleviate the conditions which produced that poverty and injustice in any country the USA deemed under its particular political sway. Since this was in the apogee of post World War II American imperialism virtually every country in the world, the representations of the USA from Korea through Vietnam were uniformly negative

Understandably though regrettably, such judgements came to include not only the nation and its ideology, but its people as well, and the rare American visitor to the GDR who was not a member of a visiting communist or leftist delegation might well find himself or herself a pariah in the country.

This meant that in 1972, when the first contacts between Brown and Rostock Universities took place in Rostock, the potential for establishing close institutional and personal ties looked dim enough to all but the most steadfast optimists.

Perhaps it is good to be able gradually to begin to forget these times. They were hard times, personally as well as in the larger political sense. The division of the country, the division of the world, the erection of walls of all kinds was an extremely painful experience for those of us who experienced it. Remembering walls is still painful, though perhaps still necessary and useful, for they are far less easily taken down than put up, however dramatic the press representations of November 9, 1989 may have appeared.

This present volume is a collection of essays and articles by those people from both sides of what was once that impermeable border between the GDR and the USA, between East and West who actively engaged in dialogue before it was common or popular, before in the minds of some it was even moral or quite legal.

The essays and articles were delivered in earlier forms at a symposium held in March, 1989 - before the "times changed" so dramatically - at Ahrenshoop, a small Baltic Sea settlement in the GDR. Some 25 people attended the symposium. Some were the veterans of almost twenty years of dialogue and cooperation. These people, some of them at least, had made the often arduous journey between Providence and Rostock many times (it could take as little as sixteen hours door to door with luck and a calm sea via Copenhagen and Gedser-Warnemuende provided, of course, you were equipped with the proper visa and documentation). The physical journeys were arduous and exciting. They were also rewarding, for then each trip brought with it proof of its own importance. In each disbelieving question "back home" - "You were where? Why? For how long? And they didn't...?" there was the immediate satisfaction of an answer which was always optimistic. "Yes, I did go, and No, they didn't..."

Each step was also a giant step even if the step itself in looking back on it was very small and apparently inconsequential. Everybody who engages in a pioneering effort knows this sensation. It produces great nostalgia when, later, as times have improved and the "path is well trodden" things seem so easy. Humans like challenge. We also like results.

The results are in very small part presented in this volume. There has been joint research, joint teaching, joint ventures of many kinds. In the past ten years since the contract was actually signed in September or 1979 by Rektor Wolfgang Brauer of Rostock and President Howard Swearer of Brown, over three hundred individuals have made more than five hundred separate visits to the partner institution. Some have been brief visits indeed and some have lasted a year or more. The visitors include students, graduate students, faculty, and administrators. All have their own memories and their own impressions of the value of the visit to themselves, their work, to the politics of the exchange itself. Most have been positive impressions. There are also a few negative ones.

The volume celebrates success. For almost two decades in very trying times, people reached out to each other and tried to make the times different. And they succeeded. A record of that success is located in every page of this volume. It is fitting too that this success should have been initiated in part by one dedicated to the study of the classics, of the "humanistic disciplines" in their most steadfast form in western academic tradition. What we learn from those "classics" is the value of thought, reflection, and the preservation of values and of life itself, if lived on behalf of that which we take to be "good".

This would have been an appropriate ending to my introduction. All of the rest is available in the essays and articles themselves.

But something happened. It often does. Just what has happened is still unclear, at least in its larger terms. The GDR has undergone a revolution or a continuation of its earlier and perhaps ongoing revolution. The West is having both a representational "field day" and lots of second thoughts. Reunification of the two German states has become a fact. The GDR has become a memory and the West, though apparently celebrating a victory over socialism, is having its own set of now less pronounced difficulties and crises ranging from drug wars to widening economic problems. The mood in the West and the East is apprehensive. The East's apprehension seems understandably focussed on "what's going to happen next?" In the West, the focus is less fixed. The apprehension more familiar perhaps? Are things going well? Will consumption really be an answer? Ecological issues are said to dominate. But ecology is a difficult term to define. Narrowly it may refer to the "greenhouse effect" and to the salvation of the whales. More broadly, it may have to do with the immediate survival of the human spirit, for that is clearly under attack and the enemy seems harder than ever to define. Perhaps it is we ourselves?

At least, the enemy is that part of ourselves which seeks to tear down those visible and external walls, like the one in Berlin but pays little or no attention to all the walls which continue to surround us - those in which we are ourselves complicit such as gender, race, and class. The Berlin wall was always good for one thing - to remind us, to help us remember - all those other walls and to work as hard with the hammer and chisel to remove them all. What strikes me about the future we now face is the powerful modes of persuasion which exist to convince us that the only enemy has been largely overcome - that enemy which built the visible wall, the one in Berlin.

We live today in a "global village." It is a place with no real geographic locality. If anywhere it is atop tall towers or in large or small studios - electronic localities which beam out to us fictional images which we now take for real. The images are the combination of dots on a screen or the oscillation of waves at a particular rate which can be encoded and decoded if one has purchased the right device.

And almost all of have purchased the right device - in some cultures like my own, that of contemporary America, we have purchased numerous such devices. Through these decoders we learn what life must be, what is happening near and far, and why it is happening and what is being done about it. The consolation of the elderly and of increasingly alienated children in western cultures and now in eastern cultures too - politically speaking, that is - the electronic imagery has persuaded us of its right to define our lives, to educate us, and to direct us.

In itself, this is no worse, perhaps, than any other form of direction and education. The content is no different from ideological domain to ideological domain. But this medium compels us to vicarious experience at a pace in a format which asks so little of us - neither concentration nor imagination - that we can and apparently do become inert beings.

Political consciousness, especially dissenting political consciousness, requires both concentration and imagination. What the content and the pace of this medium ask of us, compel in us, is the acceptance of a fictional world in place of our lives. In this fictional world, for the time being at least, and we do not know how long there will be a time to be - consumption and passivity replace the strenuous activities of concentration and imagining. The result is predictably that the actual *Tal der Ahnungslosen* -a “place” reputedly located in Dresden and other areas nearby where FRG television could not penetrate is really wherever that medium under its current ideological masters who presumably serve it as well, can reach. And that appears to be everywhere.

Television, of course, is not the enemy. It can be a powerful weapon in whatever enemy quiver it may repose. But the issue is more its impact on behalf of its current masters - and we are among them too , I suspect, more than we know - seems however pernicious indeed. As I write these words, for example, the latest Pepsi Cola commercial is playing on the television set in my study. It depicts among other things, happy people racing through the very wall through which I have labored as have all of us involved in this exchange, in this dialogue. These people are all happy and they should be, those at least who are represented by these electronic images. The final shot, the one following the close ups of the beaming children running - depicts a GDR border guard placing a flower in his uniform jacket.

These things all did happen. Children did run. People did race through the wall, laughing and happy, overcome by the events and by the suddenness of them. Border guards got flowers and put them in their uniform jackets and they, like the rest of us, experienced the sense of bewildered pleasure that the world seemed different, that real world that is.

But what I have just watched and what has ben watched by millions of viewers in this single country already hundreds of times, has nothing to do with what happened. It is a representation of these things, these joys and these sorrows, these problems and these solutions, on behalf of a company which wants to sell us its products so that its stockholders may make money - we could be among its stockholders, some of us at least.

Nothing we do, nothing we have, even our feelings, our most personal and intimate emotions, is any longer just ours or ours at all. It all exists on behalf of those who control the images.

It is this world we face after the wall, after the fall. And it is not only a world in which a problem has been solved. It is a world into which a far greater one has been unleashed.

We will have a harder time with this one. Its borders, its uncompromising walls and its border guards and customs officials are far sterner than any we have faced thus far, and its refusal to allow dialogue is absolute.

And it is just here that we again need that optimism, that steadfast faith that taking up our labors once again on behalf of something we believe to be "good", we set forth to make ourselves and perhaps by extension some others aware of this wall, this border, these guards located deep within ourselves. And that we then begin to pass through those borders, engage in those same dialogues and exchanges, confident that somehow, by dint of the effort of thought, travel, writing, and talking things will again get better, that the "good" may prevail. And as I thank all of those who have made this volume possible so too I wish us all luck, all of us who may engage in this new chapter of the ongoing endeavor.